THEY SHOULD HAVE BEEN HANGED:

War Nerd Essays on
the U.S. Civil War

Gary Brecher
"The War Nerd"

Cover image by Daniel Greenhalgh.

Contents

Introduction: How I Learned the Civil War

I t was a sad day when I outgrew my Civil War uniform. Union blue, of course. I was always pro-Union, mostly because they seemed doomed. They won in the end but they seemed doomed for years before that. So they fit well with my other heroes, the mammoths and Dire Wolves and Irish.

I really did have a Civil War uniform. A little too young to serve in the ranks (I was born in 1955), I begged for a Union Civil War uniform until my mother wrote to Aunt Molly back in Yonkers, famous for her generosity at Christmas, and got the blue tunic and pants. That was Christmas 1964, and kids who'd watched all the Civil War centennial TV shows and movies wanted those uniforms. If you look at Charlie Brown comic strips from the mid-1960s, many of them show Charlie and his proto-hippie friend Linus in Civil War kepis, either talking about San Francisco Giants baseball or having three-panel discussions about their clinical depression.

The Civil War was everywhere, and always with the same tone: it was officially a tragedy, but it was a very cool tragedy and also—this part was very important—nobody's fault. There were two sides, and most of the books and movies leaned toward the notion that it was a good thing the North won, but along the way there was always a heapin' helpin' of sympathy for the South, the gaunt ranks in grey.

This, I felt, was an outrage. I was a Union man from earliest childhood. It was not a popular position. The South had a sentimental advantage as the underdogs, and as losers. This was the 1960s, remember, and there was a lot of complacent sympathy for losers and underdogs—some of them, anyway.

I didn't see it, because I grew up around kids who seemed like direct descendants of the Confederates, and though we were friends most of the time, they scared me. And they had dads, the terrifying kind, who were as quick and lethal on the attack as any of Nathan Bedford Forrest's cavalry.

We lived in a distant suburb of San Francisco, and in the 1960s most of the families on our street were Okies who'd moved to California for jobs with the oil refineries in Martinez, just up the freeway from us. They talked about Texas and Oklahoma as their homelands, and they all had dads, I mean serious dads. My own father was a diffident, brilliant, browbeaten man. It was one day playing with my Okie friend Kenny Turnbull that I realized Okie dads were something different.

On TV, the Beverly Hillbillies were slow and sweet, and Jed Clampitt was my model of courtesy. The Okie dads on our street were non-fiction. Not cute, not harmless.

I thought dads were punchlines because my father was the punching-bag in our family, a fatally meek prodigy.

And then one afternoon I was hanging out with Kenny Turnbull in the Turnbulls' camper, looking at a dead pheasant his dad had shot on their last hunt in 'the sloughs' of the Delta. I was staring in wonder at this crazily magnificent little chicken-thing in the feathers of an emperor, when the camper door exploded open. And faster than I can tell it, Kenny's dad, Mister Turnbull, was inside with us, in that tiny space, swinging his big belt buckle at Kenny. Mister Turnbull, normally sullen and quiet, was transformed. He filled the camper with terror. He didn't say a word, and neither did Kenny. Kenny just tried to duck and cover, not very successfully. He was trapped in that camper and his dad was beating him without pause as I stumbled out, just plain terrified.

I suppose all those cans he always had were beer cans, and that beer had something to do with the rage. Or chores, or some other rule. My own parents never drank, and we expressed anger via whining or cold silences, aimed mostly at my father.

This was a new kind of dad, and it seemed connected to the fact that the Turnbulls were Okies—Confederates, in other words. So I understood in a deep, terrified way why the South kept winning even without industry or a navy or logistics. They were like Mister Turnbull. They were pure terror.

And Mister Turnbull wasn't the only violent Okie dad around. My other Okie friend Calvin invited me to his ninth birthday party. His mother, who really was a sweet woman, made a great cake (one thing about Okies, they cooked better food than we ever had) and we were all singing "Happy Birthday, dear Calvin" when Calvin's dad exploded onto the scene, swinging the inevitable belt. Calvin seemed to know what was going to happen. He took off, but his dad cornered him by the tool shed and was still flailing away at him when Calvin's mom ushered us silently away from the picnic table, out of the house, onto the driveway. We were a dim bunch but we knew that party was over.

So Confederates were never cute to me. They terrified me. Not by a long shot—I was all for the Union. The only problem was, how could the bluecoats have won? An army of Mister Turnbulls? I'd have fainted from sheer terror, facing down the Confederate coyote-yip battle cry. The notion that sane, modest New Englanders had stood up to an enemy like that was just plain implausible.

It was the way they exploded onto the scene, Calvin's and Kenny's dads. They didn't say a word, and they didn't even look like they usually did. They looked all of a sudden like drawings I'd seen of Mongol warriors. Not that they had any Asian blood—they would have turned the belt on me if I'd dared suggest such a thing—but their eyes narrowed somehow. That and the silence. Not even their kids made a sound while getting whipped. It was just the belt clanging off whatever metal was around.

Neither Calvin nor Kenny ever mentioned the beatings afterward.

But they changed. They got mean and quiet. Calvin was smart enough to be in the Gifted Program, but he didn't graduate from our miserable high school. Kenny ended our friendship by bouncing a chunk of concrete off my head, and got in a lot of fights, which was rare in a suburban high school of the hippie era—try imagining Linus and Charlie Brown in a fistfight. He vanished after graduation, supposedly up to the Redwoods to grow pot. Calvin joined the Marines. They put him on something and it zombified him pretty thoroughly the one time I saw him again. He didn't have much to say but his fists were twitching as we conversed.

Alongside the middlebrow Centennial of the Civil War, with its tone of sweet sorrow, there were other signs of the war erupting as I grew up. Places like Macon, GA and Selma, AL flared up, despite having hosted the cleansing legions of the Union. That wasn't surprising, not after my experiences with Okie dads, and neither was the sudden sprouting of George Wallace signs on our street in 1968. By then the US was messing everything up everywhere, as far as the headlines read, with this depressing "Vietnam" place proving how we could ruin anything we touched, so maybe it was just genetic. The eerie revival of names from Alabama and Georgia seemed inevitable—Dracula Was Risen from the Grave, because those stupid New Englanders couldn't hammer a stake straight.

The only anomaly is that somehow the New Englanders defeated those Okie Baptist kids, the ones who'd been beaten with belts. How was that possible? They taught us a little Emerson and Thoreau in English class, and I couldn't imagine those stolid Northern Congregationalists who seemed to spend most of their time sitting by ponds coming up with aphorisms having the reaction time to cope with such silent, instant malevolence as Kenny's dad swinging that belt. The thought of that scene still froze my blood years afterwards—Mister Austin swinging that belt-buckle like some Country and Western mace, in total silence, and Kenny dodging around the camper trying to avoid it, while I pressed myself into the wall. Mister Austin's sudden descent on that camper seemed to embody every conqueror from Jingiz Khan to Stonewall Jackson, and my own paralysis a contemptible repeat of General McClellan's big freeze at Antietam.

And yet somehow the North won. I read about its slow, painful victory in my American Heritage books and special edition Civil War Centennial books that my indulgent, upright, penniless parents bought me. I went over every one of the illustrated battles, suffering through the North's slow start and savoring its Anaconda triumph, squeezing the life out of the southern rattler.

It took me a long time to look back and figure out how it had happened. It turned out my old gods were not dead, but they weren't quite like I'd imagined them. Or, for that matter, as mainstream America imagines them. From the brilliant, furious Count Gurowski to the hollowest of hollow men, Robert E. Lee, they were weirder and more interesting than their public selves. I've tried to convey some of the wildness and sheer insane point-blank preaching of both sides. But not in a both-sides way. I've worn the Blue since Halloween 1966 and I don't pretend to be neutral.

Most of these essays were published for subscribers to the Radio War Nerd podcast. My co-host Mark Ames and I started the podcast in 2015, but it was a long time before Mark persuaded me to talk about the US Civil War. Any errors are mine alone.

Chapter One

Monitor and Merrimack, My Ironclad Gods

R *eaders should know I have my own ideological take on the Civil War. To that end, here's an invocation to the Civil War Muse.*

If there's an idea of god I could buy, it's the two ironclads, eternal opponents, fighting each other forever. That scene is burned in my brain—and not just mine, either. Real Americans see it all over the place, which is why there's a rock formation in Moab, Utah, called "Monitor and Merrimack." We see those two shapes the way Richard Dreyfuss saw Devil's Tower in the mashed potatoes in *Close Encounters*.

For you heathen who weren't raised right, the Monitor and Merrimack were the first two ironclad warships in combat. They fought to a draw in March, 1862, at Hampton Roads, a harbor near the mouth of the James River that flows down from Richmond, capital of the Confederacy. There's something permanent about their fight, though. In a way, if you believe in it like I seem to, it's like it never ended. The Monitor and the Merrimack are still fighting, and always were fighting, even before people got the idea of building them.

They're a perfect pair, because aside from the fact that no crummy wooden warship had a chance against either one, they have nothing in common. Even the way they were born showed how they were the heart and soul of the two totally different Americas that made them, the two Americas that were at each others' throats then, and still are, and always will be. I'm not saying good and evil, but...yeah, I am saying good and evil, as long as I get to admit that evil has a big pull too, I'm not automatically non-stop on the side of good. But yeah, with that in mind I'll admit: North equals good equals Monitor, and South equals evil equals Merrimack. It's like: Yay for them both, but I hope the Monitor wins.

It never does, though. The battle always comes out a draw, because that's the frustrating annoying way the world is set up. That draw is part of why the two ships fighting for eternity makes such a perfect religion. I never liked the idea of God being all-powerful because how come...well, never mind my fill-in details; everybody's got their own "If God's so nice, then how come..." to fill in.

The only way you can forgive god, or the gods, or the galaxy or whatever you call it, is if it's not all-powerful, if it's trying its best but having a very hard time, like the Army of the Potomac. And evil has its own problems and its own heroes and to be honest is pretty damn cool in its own right.

Evil has to improvise, like the South. They knew they couldn't fight the North's naval blockade ship-for-ship. The US Navy ballooned up to the biggest and strongest in the world. That's right, "strongest"—and don't you Brits give me your Royal Navy theme-music. I just wish, God how I wish, our navy had had a chance to slap you guys around circa 1862, when you were flirting with Dixie Cotton every time you thought her ex wasn't looking!

So the South had to think harder, the way besieged countries do. The Merrimack came out of the same brilliant desperation that coughed up the Me 262. It wasn't as beautiful as that interceptor, though, because it was a ship brought back from the dead. The Confederate Navy wanted to build an ironclad warship from scratch but found out they just didn't have the industrial base to make steam engines powerful enough to push all that armor plate and ordnance. So they dug up the dead US Navy steamship Merrimack, wrenched her up from the bottom of a river, cleaned her up and built a totally new superstructure for her, in-sloping walls that started with two feet of wood reinforcing, fronted by four inches of iron plate. More than anything, the reborn Merrimack looked like a WWII sub resting on the surface, with a flattened, stretched conning tower. Which is another perfect detail that can't just be an accident, because didn't the South raise her up off the mud and bring her a second life? And didn't that same South give us the first working war submarine? If you ever wanted a clue that you're dealing with the forces of the underworld (South equals under, for that matter), the history of the second life of the Merrimack would be it.

And the Underworld is always good with guns. The Confederate ironclad had gaps in the armor for 10 guns, including two big Dahlgrens, the best naval guns in the world. On March 8, 1862, the Merrimack—which the Confederates called "C. S. S. Virginia," but I'm calling it "Merrimack" because for one thing the two deities here both start with "M" and I'm not messing with something as perfect as that—sailed down the river to attack the US naval vessels anchored in Hampton Roads.

This is like the scene in every good horror film where the good people, the loyal deputies, all fire at the monster-hero without effect. The Terminator driving straight into, and I mean into, the LAPD station, that scene. The US Navy was something back then, brave, knew what they were doing—but they were in wooden ships. They didn't have a chance. The Merrimack, which had a few wooden steamships in tow like a bully's hangers-on, steamed slowly up broadside to the USS Cumberland and opened fire at point-blank range, while the Cumberland's small deck guns bounced shells off her iron sides.

Soon the Cumberland started sinking, still firing. More than a hundred sailors were dead on her decks. She went down trying to take the monster with her, just like the doomed good-guy should, because the Merrimack's iron bow ram was caught in the Cumberland's hull and Cumberland did her best to take the Terminator down with her. But it's too early in the story for the monster to die, so Merrimack broke free just as Cumberland went under.

The monster needed another victim to show its strength; you know how these scenes go. So the Merrimack turned on another wooden ship, the Congress. Congress fired back for an hour—an hour, a wooden ship firing point-blank into iron plate studded with heavy artillery! Then, sinking, she surrendered. The Confederate navy allowed the Union sailors to start evacuating until Union shore batteries fired at the Merrimack. That made the Confederate captain mad and he ordered hot shot, literally hot lead, fired on the Cumberland, which burned and sank and took more than a hundred men with her. Don't make the monster mad. It's like these people had never seen a horror movie, which they hadn't.

The third time, something different happens. That's in every story I know. And that's how it was this time. The Merrimack looks around for another target, finds the USS Minnesota. And let's stop there.

Minnesota—the state, that is—was the purest of the pure back then. The furthest north, the furthest west, the two good directions. A new state, no slaves, settled by innocent Scandinavians who never really got the Ulster crazy hum America always carried in its belly. And now this iron monster from the murkiest heat of Dixie turns on the ship named after the headwaters, where the Mississippi is still a clear little brook. It's too perfect. Tell me this isn't a battle of gods!

The Minnesota ran aground, like the leading lady tripping over her high heels as the monster bears down on her. But since the Merrimack, loaded down with all that heavy ordnance and iron plate, had a much deeper draft, it couldn't even get close enough to Minnesota to kill her. So it had to draw back, sulking like Christine when some victim makes it across the ped-xing, headlights just aching to plow through the cheerleader.

Now it's night. The Merrimack steams home for re-pairs—Schwartzenegger digging buckshot out of his hydraulic arm in that skid row rented room—while the rest of the US Navy waits to be destroyed, one wooden ship at a time.

A second ironclad was coming toward Hampton Roads, the equal and opposite, the good twin, as pure a product of the good, well-fed North as Merrimack was of the feral South: the Monitor, a spick and span tin can floating a big gun. Ahura Mazdā to the rescue against the ironclad Ahriman. In the Mayan religion, the hero is, or are, twins. I've always liked that idea, but not identical twins, more like opposite twins, but still twins, more like each other than they are like any of the ordinary mortals they fight for/against.

The Monitor was born in cool, rational heads, the opposite of the South's booby-trap rigging desperation. The Monitor plans were drawn up on clean paper in the offices of brilliant, educated men. It was the child of an official body called "The Monitor Board," ap-

pointed by Gideon Welles, Secretary of the Navy, a typical Victorian superman, and designed by John Ericcson, a Swedish engineering
genius. Minnesota, Ericcson...I'm telling you, this is about the most
Northerly north meeting the most Southerly south, not even geogra-
phy, this is a god-fight.

The North had the immigrants' brains, and immigrant desperation
is a match for the cool Satanic inventions of the South any day. Ericcson's design was as pure, clean, and strange as the Merrimack was
ugly, dark and lethal. The turret was a perfect cylinder, the engine and
hull were smooth minimal steel, everything had the almost-alien look
of a truly brilliant design. Ericcson's model was called a "Monitor"
because it was like one huge eye, an armored cyclops. Or, like other
people with no respect for religion said, it looked like a tin can on
a board. The tin-can turret held two huge guns, 11-inch Dahlgrens.
That was all that showed above water. None of the fuss and mess of
other ship architecture at all. It was the cleanest design in the history of
weaponry. Samurai swords look over-embroidered and busy compared
to the Monitor.

And naturally it looked helpless against the big, bad, grinding mass
of the Merrimack. The contrast is everything here. It's the basis of my
religion, if you want to put it that way. They're equal, more or less, but
they don't look equal. The Monitor looks too small and fragile, like
it should. But it's faster, it's smarter—it's David to the Merrimack's
Goliath, only better, because that Bible story never worked for me
thanks to Yahweh already declaring for David. Once an all-powerful
god announces he's on your side, where's the suspense, where's the
heroism? There is none. David could have picked a piece of lint off
his tunic and blown it toward Goliath and it would have blinded him
and made him trip and break his neck. The rock was just a prop.
That's cheap, and dirty—who's the bully anyway, Yahweh, you cheat-

ing punk up in your cloud, fixing the fight to get good odds on the little guy? That's why Zoroastrianism seems like a better explanation for the way things are: Good has to fight, and sometimes it loses. But it always has to fight.

The Monitor, I keep trying to say, is a different religion, a Union religion that Yahweh would've had a jealous little-girl fit over. It had nothing on its side but the cool Northern brains that dreamed it up, and they hadn't had much luck against the hot crazy demons of the South in the first year of the war. It must've been some seriously cold comfort for the wooden ships of the Navy to see their tuna-can savior stationing itself just offshore of the Minnesota. "Oh, great: My bodyguard, the five-foot nothing science nerd!"

When the Merrimack came out of its upriver lair at dawn on March 9 and steamed downstream to finish off the Minnesota, the Confederates didn't even recognize the Monitor as an opposition vessel. They thought it was a spare boiler being towed across the harbor. Then the boiler opened fire with its 11-inch Dahlgren, bigger than any gun the Merrimack had. The Monitor was big in offense, but a tiny target, the design goal of any armored vehicle. In fact, if those 11-inch guns had been charged with a full load of powder they could have pierced the Merrimack's hull. But the Northern design team was a little wary of that much bang in a tiny turret, and I can't say I blame them. So the Monitor's big shells dented but couldn't break the Merrimack's iron plating.

Merrimack fired back with a broadside that missed Monitor's little turret but hit the Minnesota, the big TKO'd Swede Monitor was protecting. The Minnesota, game as ever, fired its cannon back, stuck there on its mudbank. A big dumb jock, but brave, and on the right side. You have to love the Minnesota. And Cumberland, for that matter. That's what's better about this religion: everybody's got a good job to do, and even the evil is great.

They fired at each other point-blank, with the Monitor's big guns in their 360-degree rotating turret always bearing directly on the Merrimack, and the Merrimack's big black slab-sides absorbing all that punishment and still spitting back shells. If the Merrimack had had solid shot for its Dahlgrens, they might have zipped right through the Monitor's thinner armor plate. But the Underworld always has to improvise, it's always under embargo like Dixie was, and they didn't have solid shot in the right size, so the Merrimack had to be content with exploding shells.

Finally one of them blew just in front of the viewing slit of the Monitor's turret and the little savior withdrew. The Merrimack was happy to call it off too, seriously dinged up, its zombie engines, that'd lain in the mud a long time before being resurrected, grumbling and threatening to quit, the armor plate buckled and bent.

Both deities steamed away and both declared victory. And both those particular ships died, in another perfect-opposites way: the Merrimack died by fire, burned by her crew when her home port was captured by the Union, and the Monitor died by water, sunk when high seas slopped over her little turrets out on the ocean.

But those were just the two temporary things that held the fight that day. North and South, Minnesota and Virginia, they're still fighting. Not just in DC but my head. And in Moab. Everywhere, actually, because there are other pairs like that. For me, making up a better religion while I had to sit in the pew and listen to the grownups ranting, the Monitor and the Merrimack were just the first and best of the fighting pairs. The next I recognized was the same one every normal person has in their heads: the whale and the squid, fighting forever in your head and in the black depths of the ocean.

The Merrimack comes up from the mud to fight the Monitor; the whale goes down to the underworld to fight the squid. And just like the Monitor and the Merrimack are both perfect and wonderful, the whale and the squid are both great too, the muscle, big lungs and strong jaw of the whale against the cold boneless pull of the tentacles. They're not exactly equal; I'm always on the side of the Monitor and the whale. But I worship the Merrimack and the squid too, where they belong, proper gods of the underworld.

And they fight forever, amen.

Chapter Two

The Confederates Who Should've Been Hanged

*N*ow I've confessed my Manichean leanings, but let's not get carried away. This is a pro-Union book and I've got no patience with those Dixie simps out of the Confederate brass. What they needed was a good, sturdy gallows.

"The problem of [the] war consists of the awful fact that the present class of men who rule the South must be killed outright..." *Sherman Letter to General Sheridan on the eve of the March to Atlanta*

It's a tricky question: Which representatives of Southern manhood should have danced in the air, come April 1865?

I think we can all agree, Lost Cause loons aside, that every Southern officer was a traitor who'd earned the right to dangle. But alas, if you go hanging every single officer of a defeated army, you can very easily end up with a nasty insurgency on your hands. Those men had brothers and cousins, quite a lot of them. In the 19th-century Anglo world, wealthy families had lots of kids, and their brats tended to survive at a higher rate than starved poor kids. Take the ex-Confederates I'm going to talk about here: Jubal Early and Porter Alexander each had nine brothers and sisters, and Nathan Bedford Forrest had 11 siblings. Kill the twenty-something son who served with the Confederacy, and you have to deal with four or five kid brothers swearing revenge.

At the same time, it's clear that the policy the Union actually pursued—not hanging any Southern officers except the miserable wretch who commanded Andersonville POW camp—failed miserably. A decade after we defeated the Confederacy at the cost of 300,000 loyal Union soldiers' lives, the same planter oligarchy was running the South again, terrorizing the Freedmen and women who were our only loyal allies during the war, making sure black people never got a chance to vote, running them off their farms, doing their best to recreate slavery without the name.

And it might have been possible to prevent that disaster by hanging key ex-Confederate officers in the spring of 1865. All the leaders of the post-war terrorist fascist gangs that disenfranchised African Americans in the South were former Confederate officers. If we'd thinned their ranks in an intelligent way, Reconstruction might have been something other than a grotesque and bloody farce.

There are some obvious guidelines for thinning the ranks of a dangerous group at their moment of defeat:

•You don't kill the figureheads. They become martyrs quickly, and they've usually passed their peak.

•You don't kill incompetents. Keep them alive as long as possible.

•You don't kill the corrupt. You buy them and use them to turn your former enemies against each other.

•You kill the exceptional, the most ruthless, fearless, unkillable leaders. If you don't, you'll regret it.

I'm not talking about justice here. Justice would have demanded hanging every Confederate with a rank of colonel or higher. But often the higher the rank, the older the man, the more tired and harmless he was by the time of the Surrender. Jefferson Davis, for example; justice says he should have hanged, if not tortured to death, but Davis was such a disaster as a Confederate symbol that he wasn't one of the more dangerous post-war figures. Better to let losers like Davis live on as buffoons rather than kill them and start the songs and poems going.

No, I'm talking about practical killing. Who were the most dangerous ex-Confederates in 1865? Could they have been identified and killed before they neutralized all the gains of America's most costly war?

You can assume that in a group as big, as tough, and as dispersed as the Confederate officer corps and its core civilian elite, there will be a huge range of reactions to surrender. Some will commit suicide, like the South Carolina long-haired fanatic Edmund Ruffin did in June, 1865. (He'd planned to do it on April 9, but he had company that day, and as a polite Southern host, he was forced to live on another two months before putting the barrel of a rifle in his mouth.)

Others, like Lee's very talented artillery officer Porter Alexander, will be drawn to guerrilla warfare.

This will have particular appeal for younger officers, and those (like the buffoon-ish Sterling Price) who are far from the main front and can't grasp the reasons for the defeat. Price had the brilliant idea of fleeing to Mexico to take service under Emperor Maximillian, soon to be known as "that dark spot on the pockmarked wall." Price came back to Missouri and died, which was by far the best thing he ever did.

None of these men, or even more effective postwar irregulars/bandits like the James Brothers, ever represented a real threat to the Union victory. That threat came from ex-Confederate officers who were cold-blooded and intelligent enough to bide their time, take advantage of the North's ridiculous leniency, and form quasi-legal organizations to negate every gain for which those 300,000 soldiers died. These were the men who needed to hang in April 1865.

It's easy to identify the two ex-Confederate leaders who did the most to ruin the lives of the African American and poor-white Southerners after the war: Nathan Bedford Forrest in the West, and Wade Hampton in the East. If those two had been hanged in 1865, American history might have gone in a different direction, and frankly, almost any outcome would have been better than the debacle that actually followed the war.

Nathan Bedford Forrest, un-hanged, went on to front for a little group you may have heard of, called the KKK. Wade Hampton, who gets less press but was probably the worse of these two monsters (admittedly, it's a tough competition) created America's first homegrown fascist group, the Red Shirts, and used them to terrorize black voters, ensuring his election as South Carolina's first post-war racist senator in 1876.

And these guys didn't suddenly turn bad after the war. Both of them were born bad, and had done enough during the war to deserve death by any moral or legal criteria you care to name.

Forrest was a slaver and a killer long before the war, but he distinguished himself among the bloody Southern officer corps by his fondness for "No Quarter" orders. "No Quarter" was much more common in the Southwestern theatre of the war than most people realize. The James brothers, Quantrill, Anderson—those guys didn't come out of nowhere. They were typical of the Southern irregular cavalry, and Forrest was the best, most ruthless leader they had. Forrest didn't like taking prisoners; he preferred killing them on the spot. And it worked for him, once his rep got around. Many weak commanders surrendered to him rather than face the prospect of being slaughtered if he won.

When he attacked Fort Pillow in April 1864, Forrest encountered a garrison that wouldn't surrender, and was half African American. The black troops were from two artillery units, backed up by raw infantry. Forrest's raiders outnumbered them 1,500 to 600 and Forrest expected to win easily. He issued one of his standard threats after initial skirmishing, telling the Union commander he and his men had fought well enough to be "entitled" to be treated as POWs if they surrendered, but if Forrest was "forced" to attack, he couldn't guarantee their safety.

It worked, many times, but it didn't work on the second-in-command at Fort Pillow, who replied, "I will not surrender." Forrest's men overran the fort and killed every black soldier they could find. One of the Confederates who took part in the massacre reported it like this:

> "Words cannot describe the scene. The poor deluded negroes would run up to our men fall upon their knees and with uplifted hands scream for mercy but they were ordered to their feet and then shot down. The whitte [sic] men fared but little better. Their fort turned out to be a great slaughter pen."[1]

---

1. "The Racial Atrocity of the Civil War: The Massacre at Fort Pillow" by Kari Foley. Chronos, Vol. 1, 2007 (pp.40-52), p.48.

After a half hour of slaughter, Forrest resumed command, and sent a proud dispatch boasting that the "river was dyed red" with the blood of the African American soldiers. Forrest was a master of terror in war, and saw the massacre as a good way to neutralize the growing number of African American soldiers the Union was recruiting. He wrote, using the modest passive mode, "It is hoped that these facts will demonstrate to the Northern people that negro soldiers cannot cope with Southerners."

Forrest later realized he might have gone too far for his own safety and started backpedaling. In a less than coincidental incident, Bradford, the Union commander who'd witnessed the whole massacre, was shot "while trying to escape" from Forrest's men.

So by the time of Lee's surrender, Nathan Bedford Forrest was guilty of murder several hundred times over. He was kill-able. He was the most eminently kill-able man who ever lived. He deserved death many times over. But he was allowed to return to civilian life, which for him meant becoming the First Grand Wizard of the KKK. And please, don't go on about how he "later renounced the violence of the Klan." What Forrest didn't like about the evolution of the KKK was that he, Forrest, wasn't in complete command of it, and that he felt its violence was amateurish. He was a pro, and he wanted artistic control over the symbolic violence in which the Klan traded.

Forrest's survival after the war was a disaster on any level you want—legal, moral, political. Nathan Bedford Forrest should have graced a gallows in the spring of 1865, and that should have been clear at the time to any resolute Union government.

Wade Hampton, the other leading candidate for a spring hanging in the wake of Apomattox, was, if you can believe it, even worse than Forrest. At least Forrest was a self-made monster; Hampton was a rich boy, the son of the South's leading slave-holder. Hampton's family owned more than 3,000 human beings, but rich didn't mean pampered in the planters' world. Wade Hampton III was not pampered. His hobby was hunting bears. With a knife. None of these guys were pampered. In fact, you feel a lot fonder of pampered, soft people after reading about these monsters.

Like a lot of tough kids, Wade Hampton III had something to prove. His grandfather, the original Wade Hampton, was the commanding American officer at the Battle of the Chateauguay in 1813, against a small, hastily assembled force of Mohawk Indians and Canadian militia. You don't hear much about that fight in America, just as you don't hear much about the Battle of Patay in Britain. If there's one thing us Anglos are good at, it's burying our humiliations. Chateauguay was a complete humiliation, with an American force routed by a mixed militia half its size, then lost in the woods by Wade's grandfather.

By the time the Civil War started, Wade Hampton III was 42 years old, with no military experience. But he was a mean bastard, he knew how to ride and kill, he was willing to use his own money to raise his own "legion," and he rose fast. In fact, one of the best ways to identify candidates for hanging is to look at fast risers.

In the whole Confederate army, only two men who started with no previous military experience rose to the rank of Lt. General: Wade Hampton III and Nathan Bedford Forrest. That's a good noose-fitting device right there.

And if you're looking for good legal cause to hang ol' Wade, you won't have much work to find it. Hampton talked his head off to Sherman's officers, late in the war, as they arranged the surrender of Johnston's forces, and his main theme, as recorded in multiple Union officers' memoirs, is shooting deserters and "recruiting" new troops at gunpoint. Military life, for Hampton and many another Confederate officer in the last year of the war, consisted of rounding up deserters, shooting every one who didn't seem useful, and re-enlisting the rest by holding a pistol at their head until they sang "Dixie" in the proper key. There's no knowing how many Union men Hampton killed, but he boasted about killing dozens of reluctant Confederates.

Hampton survived the war, alas, in the same state of mind of most of the planters: not having learned a damned thing except to hate Yankees, African Americans, and anyone else who failed to genuflect to the Lost Cause myth that his buddy Jubal Early was peddling—the South's version of the ol' "stabbed in the back" myth so popular with certain Teutonic parties of the 1920s and 30s.

As the North lost the will to enforce basic human rights for African Americans and white dissidents across the South, Hampton made his move to regain control of South Carolina for the planter elite. He borrowed an honorable symbol, the "Red Shirts" of Garibaldi's insurgents, and made the red shirt the mark of his own racist militia. The South Carolina version of the Red Shirts murdered African American leaders (150 of them during the 1876 Senate election, by one account) terrorized black voters and white Republicans (yeah, the Republicans were the good guys in those days) from voting, and indulged in any private violence that happened to interest its members. The 1876 election, with Hampton vs. a Reconstructionist, was a bloody draw, but Hampton's fascists wanted it more and he eventually simply took power. He never looked back, and neither did South Carolina. Any threat of a new South, where something other than class or money might determine your chance in life, was wiped out for a century.

An outcome like that is worth preventing. If a few hangings had interrupted the premature love-fest between (white) North and (white) South in 1865, that outcome might have been avoided. And it would not have been difficult to identify the Confederate leaders most likely to organize treasonous groups like the Red Shirts and KKK. Both were led by civilians who rose quickly through the ranks, ending up as Lt. Generals—the only two men to follow that trajectory in the whole huge Confederate army. Both these leaders, Forrest and Hampton, were notable for their efficiency and extreme brutality throughout the war. Both were relatively young. Both were unrepentant racists and secessionists. For all these reasons, they were all obvious candidates for the top spots on a gallows list.

Granted, it might not have been possible to isolate their names among other brutal, successful, young, civilian-origin leaders. But there's a simple solution for that problem: Hang every damn traitor who fit that bill.

Chapter Three

The Diary of Adam Gurowski

E *ven now I can hear people saying, "No one thought that way back in 1861." The hell they didn't. We've just refused to listen to those who did, dismissing them as fringe voices, if not lunatics. The Polish exile Adam Gurowski knew exactly what had to be done and exactly how to do it but was, of course, ignored by the "wiseacres" in Washington.*

Every real American is a Civil War fan, and that's the problem. We're sentimental about it, convinced it was an unavoidable tragedy and nobody could have solved it with less bloodshed than it took. This is just plain wrong. There were plenty of people who knew the Union needed to clamp down on the South Carolina lunatics before their madness spread. And said so at the time. The most clear-headed of these prescient commentators was an expat Polish Count, Adam Gurowski, who, while working as a translator in Washington, DC, screeched to his diary[1] that the Lincoln administration's policy of appeasement was going to lead to a much more massive, costly, gory war.

Nobody listened to him, of course. Gurowski is considered to this day as a lunatic or, at best, a marginal gadfly simply because he doesn't share our delusion that the Civil War was an unavoidable tragedy. He saw it more as a fire: if you put out the fire, problem solved. If you don't, you've got trouble.

Gurowski came by his knowledge of insurgency the hard way. He took part in the Polish uprising against Tsarist Russia in 1830 and then fled for his life when it was crushed. He ended up in Paris, joining many progressive factions, tried returning to Poland, found that the Tsarist regime had not mellowed, and then came to the US in 1849. In 1860, he was in Washington, DC, working as a translator and diplomatic advisor to the Union and tearing out what little remained of his hair as he saw the grovelling Buchanan administration give way to an even more grovelling Lincoln cabinet, who seemed convinced that if only they asked the slave owners really, really nicely, they could be sweet-talked into avoiding the war.

Gurowski had seen real revolution and knew what contemptible nonsense this was. And, being a Slavic intellectual, he said so over and over with perfect clarity but perhaps imperfect tact. And, as Lincoln's craven 1861 policies show, Americans value tact over just about anything, including saving the lives of hundreds of thousands of soldiers by the timely suppression of a slave-owner conspiracy.

---

1. *Diary...: From March 4, 1861, to November 12, 1862* by Adam De Gurowski (Count) (HardPress 2018) ebook.

Gurowski could forgive anything except the slave owners of the South. He shrugged off his own impoverishment and exile at the hands of the Tsarist regime, describing Tsarist Russia as a reliable friend of the United States against the South. The only people he really hated were the Southern slave owners. And that hate has absolutely nothing personal to it; those Dixiecrats hadn't done Gurowski any harm. In fact, he talks constantly about their fancy courtesies, which worked all too well on other European observers—but not on Gurowski. He mixed with a much more radical set, which valued liberating the slaves more than the courtesies of the drawing room.

We have a kind of half-assed idea that people back in 1861 were a little slow. You know: Brave as Hell, and with a steely simplicity we can't match, but too primitive to get stuff like, well, the racial issues and such. Wrong. That might apply to most Americans in 1861, but that was only because we were a little out of touch with Europe, and I mean "Europe other than England," because England in 1861 was a hardcore reactionary power, the most evil, far-right country in Europe, and getting our ideas from there meant we were tapping a tainted source.

Gurowski, coming out of a more wholesome, Parisian tradition, won't even use the common terms for our slaves. He calls them "Africo-Americans," as if he almost anticipated we'd end up considering "African American" the acceptable word. And he has pure contempt for any talk that there's some deep, dangerous difference between them and all the other hyphenated American crowd. We take it for granted that Lincoln's cautious tiptoe around race was inevitable and necessary, but to Gurowski, it's just another product of Lincoln's provincial ignorance. Gurowski furiously told his diary in August, 1862:

"In his interview with a deputation composed of
Africo-Americans, Mr. Lincoln rehearsed all the
clap-trap concerning the races, the incompatibility to
live together, and other like bosh... I pity Mr. Lincoln;
his honesty and unfamiliarity with human affairs,
with history, with laws, and with other like etceteras,
continually involve him in unnecessary scrapes."

Gurowski didn't love the slaves in a pitying, Bostonian way. He
considered them simply an oppressed peasantry, like German or
Ukrainian serfs. In fact, he thinks they're better material than their
so-called masters:

"Men like this [Jefferson] Davis... roar against the
African race. The more I see of this doomed people,
the more I am convinced of their intrinsic superiority
over all their white revilers, above all, over this slave-
holding generation, rotten, as it is, to the core. When
emancipated, the Africo-Americans in immense majority will at once make quiet, orderly, laborious, intelligent, and free cultivators, or, to use European
language, an excellent peasantry; when ninety-nine
one-hundredths of slaveholders, either rebels or thus
called loyal, altogether considered, as human beings...
constitute caricatures and monsters of civilization."
(March, 1862)

That was his view of the future of black Americans: give them
land, like any European peasantry, and they'll work it happily, gaining

wealth slowly and cautiously. That was their dream, too: that's where the cruel postwar Dixie joke about "40 acres and a mule" comes from. He's explicit about the improvement in Southern civilization that will take place once the land is distributed among the freed slaves:

> "The Africo-Americans are the true producers of the Southern wealth—cotton, rice, tobacco, etc. When emancipated and transformed into small farmers, these laborious men will increase and ameliorate the culture of the land; and they will produce by far more when the white shams and drones shall be taken out of their way. In the South, bristling with Africo-American villages, will almost disappear fil-libusterism, murder, and the bowie knife, and other supreme manifestations of Southern chivalrous high-breeding." *(March, 1862)*

Notice the bluntness about what will happen to the traitors of the Confederacy: They'll be "taken out of the way." One of the most surprising things about reading Gurowski as an American is his willingness to be hard on the people who started the whole mess, the white Southern planters and their idiot dupes in the ranks. That's something you never, ever see in domestic Civil War stories. And Gurowski's toughness here makes a stunning sense in context because, as he keeps trying to tell us, there's gonna be a lot of pain from a giant war like this, one way or another, so for God's sake why not inflict it on those who deserve it?

Gurowski rails at Lincoln's unwillingness to hurt the feelings of highly placed incompetents like the commanders of the Army of the

Potomac in the same way he grits his teeth at the administration's dread of offending the traitors in Charleston.

You see this squeamishness about hurting the feelings of highly placed incompetents still going strong in contemporary America. Americans who went through the Iraq War were much more stoical about the hundreds of thousands of lives destroyed than they were about the fact that some malcontents hurt George W. Bush's feelings with their protests.

In this way it's kinda interesting that Gurowski's favorite American civilian isn't our hero Lincoln (he thinks Lincoln is a sad case, a nice enough guy but out of his depth and soft on treason) but Edwin Stanton, who's usually the villain in our domestic accounts. And why is Stanton a bad guy? Well, he hurt people's feelings. Which may be why Gurowski calls him the "last Roman" in America.

> "When all around me I witness this revolting want of energy,—Stanton excepted,—this vacillation, these tricks and double-dealings in the governmental spheres, then I wish myself far off in Europe; but when I consider this great people outside of the governmental spheres, then I am proud to be one of the people, and shall stay and fall with them." (*March, 1862*)

For Stanton, there is one goal, and that's to win the war. Everything and everyone else can go straight to Hell. That's a little harsh for the James McPhersons and the Bruce Cattons, but it's music to Gurowski, who's sick of provincial manners, provincial intellectuals, and provincial "strategy."

The finest passages in Gurowski's journal are the ones in which he tears apart the notions of "strategy" that dominated Federal military thinking before Grant and Sherman took over. After the battle of the ironclads, Gurowski, who'd been begging for the occupation of Norfolk since the start of the war, said:

> "If Norfolk had been taken months ago, then the rebels could not have constructed the Merrimac. Norfolk could have been easily taken any day during the last six months, but for strategy and the maturing of great plans! These are the sacramental words more current now than ever. Oh good-natured American people! how little is necessary to humbug thee!" (*March, 1862*)

In Gurowski's view, the word "strategy," from the mouths of idiots like Winfield Scott and George McClellan, meant delay, piling up of superfluous artillery and reserves—any excuse to avoid just going in, burning and hanging the insurgent elite in its dens in Richmond and Charleston, and dismantling the rebellion before it had time to organize itself. Over and over, he brings a hard, cold, European veteran's take to all the puffery from the amateurs running the Army of the Potomac. And after a brief honeymoon, when he thought that anybody would be an improvement on Scott, he sours on McClellan, who he actually calls "McNapoleon." Let us pause for a moment and acknowledge this as the best insulting military nickname in history.

Gurowski had actually been on those European battlefields and seen what was involved in suppressing a rebellion, as in he'd had it done to him and his cousins and his friends by the Russian army.

His Napoleon wasn't the post hoc version West Pointers got out of
Hardee's translation of Jomini but the officer who made his bones
blasting royalist rioters with artillery in the streets of Paris. He knew
that it's a matter of quick brutality—grapeshot, the noose, and the
bayonet applied in a timely fashion. The very last thing you do with
a rebellion is give it time to simmer, try to suck up to it, and ask
it if it can be placated. But that's what he had to watch Lincoln,
McClellan, Buell, and Scott do, until the straightforward commanders
lucky enough to be far from DC took over. For Gurowski, "strategy"
becomes a pejorative. He grinds his teeth through his entries for the
bad years, 1861-62. You can hear the rage every time he has to spit out
that word:

> "Strategy—strategy repeats now every imbecile, and
> military fuss covers its ignorance by that sacramental
> word. Scott cannot have in view the destruction of the
> rebels. Not even the Austrian Aulic Council imagined
> a strategy combined and stretching through several
> thousands of miles.... The people's strategy is best: to
> rush in masses on Richmond; to take it now, when the
> enemy is there in comparatively small numbers. Rich-
> mond taken, Norfolk and the lost guns at once will
> be recovered. So speaks the people, and they are right;
> here among the wiseacres not one understands the
> superiority of the people over his own little brains."
> (June, 1861)

You'll notice that Gurowski is so angry here at the dotard Scott
that he uses the worst insult any European military commentator can

think of: He compares him to the Austrians. Notice also that he says the ordinary American is a better military theorist than the half-bright West Point products running the war. He doesn't say this as a way of pandering to "the common man." He's the ultimate intellectual and has a deep respect for true professionals. But he understood long before anyone in DC did that a regional rebellion has to be crushed quickly and mercilessly, or it will turn into a long and disastrous war. It wasn't until Sherman finally figured that out and acted on it that the war was decided.

Gurowski had seen all-out war, practiced on Poland, and knew this war would come to that. For him, "strategy" becomes a euphemism for something like cowardice, as practiced by provincial military pedants who don't have a clue what it means to be wholly engaged with anything. He says this very clearly, long before anyone else got it:

> "McClellan is ignorant of the great, unique rule for all affairs and undertakings,—it is to throw the whole man in one thing at one time. It is the same in the camp as in the study, for a captain as for a lawyer, the savant, and the scholar." (*January 1862*)

Gurowski didn't know Nathan Bedford Forrest, but if he had, he would have endorsed Forrest's view totally: "Get there first with the most men." It's the opposite of what McClellan meant by "strategy," but it would have worked in 1861 in a way that the Baldrick-level cunning of McClellan and Scott did not.

Mainstream historians always make excuses for the federal government circa 1860 by pointing out that military intervention would have been almost impossible. The regular US Army was a tiny force dispersed over dozens of barracks throughout the vast extent of the US. But they never seem to note that they were the same conditions that prevailed in 1832, when Andrew Jackson simply threatened to invade South Carolina if they dared to withdraw from the Union of the states. Jackson didn't have a vast standing army either, but nobody who knew him had any doubt that he was quite capable of raising a fearsome ad hoc force and levelling the mansions of Charleston. He'd made his abilities quite clear in the Siege of New Orleans in 1814, where he assembled an army composed of some of the most despised groups in early 19th-century America—backwoods men, ex-slaves, Native Americans, and Cajun pirates—and slaughtered well-trained, massive force of British regulars. Of course, no one ever compared James Buchanan, the pitiful lame-duck American president in 1860, to Andrew Jackson except, perhaps as a punchline. The idea of James Buchanan threatening Charleston SC is absurd, not only because Buchanan worshipped the planters in some bizarre, grovelling way, but because he was clearly far too feeble and slow to raise any sort of military force. But if there had been any other American politician in office in 1860, it would have been very easy to construct a strategy bypassing the regular army, drawing on the infuriated patriotism of the upper-Atlantic and New England states and equipping a few thousand recruits quickly enough to level Charleston, if not burn it to the ground, before the Confederacy could get started. Under Buchanan, of course, that could not happen. But let's be clear: the problem was not the logistics of a tiny dispersed federal army or some absurd Constitutional quibble. Jackson had had no time for any quibbles. His statement was simple: Secession was treason and would be punished accordingly.

Gurowski, coming from a long and bruising experience in European militarized parliamentary crises, can't even bear to entertain the notion that a Constitutional quibble would give anyone a moment's pause. He had experienced the law of force as the final arbiter of European controversies in very painful, direct ways, so he soon realized that most American commentators who babbled about what could be done under the Constitution had no sense of what must happen in an emergency override. He may not have even been aware of Jackson's Secession Crisis, but he knew that Jackson's course was the correct one:

> "Most of the thus-called well-informed Americans rather skim over than thoroughly study history. Above all, it applies to the general history of the Christian era, and of our great epoch (from the second half of the 18th century). Most of the Americans are only very superficially familiar with the history of continental Europe, or know it only by its contact with the history of England. Many of them are more familiar with the classical wars of Alexander, Hannibal, Cæsar, etc., than with those of Gustavus, Frederick II, and even of Napoleon. Were it otherwise, strategy would not to such an extent have taken hold of their brains." *(January, 1862)*

What he means is not just that the technique of making war has changed since the Classical era but that all wars get cleaned up as they age in our memory, getting more chess-like and less bloody as they're copied and re-copied into chapters to be studied. If you're reading about Hannibal in your study in New York or Pennsylvania, it all looks very neat and clever. If you're remembering what it was like to flee Poland with columns of smoke showing where the Russians were advancing, you have a much more realistic take on war.

European war was never the clean business a lot of American war buffs consider it. The problem is that we learn about it from the British, cuz we're too goddamn lazy to learn any other languages. And the Brits had one huge advantage: They fought their wars on somebody else's land. Somebody else's village got burned, somebody else's sister got raped and bayoneted. Which was almost like nobody's house got burned, nobody's sister was impaled.

Not that Gurowski writes out of personal bitterness. He doesn't take this personally at all. That's what's so alien and magnificent about him. His horrific experiences in Poland simply helped him understand what would have to be done sooner or later to crush the slave owners. But his experience doesn't prejudice him in any way. He's not sentimental about Poland; if anything, he's anti-Polish and very definitely anti-Catholic, disgusted with the Irish-Americans' pro-Southern leanings, he says:

> "The pro-Romanist clergy is more furiously and savagely pro-slavery than are the Rhetts, the Yanceys, in the South; the poor Africo-Americans are, if not the truest Christians in this country, at any rate their Christianity is sublime when compared with the pro-Romanism." *(November, 1862)*

In everything he writes, Gurowski shows how totally he's passed beyond the personal, tribal, local consideration. We don't even aspire to that sort of objectivity now. His first doubts about McClellan, who, he figured, had to be better than the quasi-traitor Scott, come when McClellan appoints a relative to an important job:

> "McClellan makes his father-in-law, a man of very sec-
> ondary capacity, the chief of the staff of the army. It
> seems that McClellan ignores what a highly responsi-
> ble position it is, and what a special and transcendent
> capacity must be that of a chief of the staff—the more
> so when of an army of several hundreds of thousands."
> *(September, 1861)*

The whole idea of favoring a mere relative when so much is at stake seems to shock him:

> "American nepotism puts to shame the one practised
> in Europe. All around here they keep offices in pairs,
> father and son. So McClellan has a father-in-law as
> chief of the staff, a brother as aid, and then various
> relations, clerks, etc., etc., and the same in some other
> branches of the administration."[2]

Reading Gurowski, you realize that "nepotism" and "feelings" are connected; they're both part of the personal, familial, ethnic, national series of concentric circles that Gurowski has stepped through in his own life—very painfully, at very great cost—and now expects Americans, who he considers "the best people in the world," to transcend

---

2. March 1862

too. I'm afraid we kind of let him down, there. And he seems to know
it, now and then, in his diary:

> "Now, for the first time in my life, I realize why, during
> the last stages of the dissolution of the Roman em-
> pire, honest men escaped into monasteries, or why,
> at certain epochs of the great French revolution, the
> best men went to the army.... Ah! to witness here the
> meanest egotism, imbecility, and intrigue, coolly, one
> by one, destroy the honor and the future of this noble
> people. Curse upon my old age! above all, curse upon
> my obesity! Curse upon my poverty! What a cesspool!
> what a mire! Only legal slaughterers all around! O,
> could I go to a camp! but, of course, not to one under
> McClellan." *(October 26, 1862)*

Gurowski's call for a simple, blunt, ideological war had a few American adherents, like Thaddeus Stevens, but we've managed to forget
most of them; and even the best of them were tainted, from Gurowski's perspective, with an ignorant, largely unconscious white nationalism—which, like the rest of their fallacies, comes from having a little
learning, rather than honest ignorance or true erudition:

> "If those would-be knowing arguers on slavery, race,
> etc., were only aware of the fact that such people as the
> primitive Greeks, or the ancestors of classical Greeks,
> that the ancestors of the Latins, that even the roving,
> robbing ancestors of the Anglo Saxons, in some way
> or other, have been anthropophagi, and worshipped

fetishes; and even as thus called already civilized, they sacrificed men to gods,—could our great pro-slavers know all this, they would be more decent in their ignorant assertions, and not, so self-satisfied, strut about in their dark ignorance." *(January, 1862)*

That line from Pope about the dangers of "a little learning" could be Gurowski's epitaph for America. There was something great here, out among the honestly ignorant people, he thought, but it was spoiled, like the war itself, by going halfway—halfway toward equality, halfway toward true learning, halfway to crushing treason. He wished we were hot or cold, for we were, and are, kinda lukewarm. Lukewarm and sentimental, nostalgic and parochial.

From Edmund Wilson[3] on, we've actually fostered the belittling of the world back into those old categories, those circles of autobiography, family history, ethnic identity, and sentimental nationalism. Hell, we've even added gender identity, and the only one we've subtracted, class identity, is the only one that Gurowski would have endorsed. It's a roller-coaster of a read, this magnificent dinosaur's diary. First he takes you up for a little while into the world you can sense in those late Victorian Gods, the world that might have been, but over and over you follow him down the big dipper into the petty world of local loyalties and sentiments, "feelings" and cheating, that ended up winning the peace.

---

3. Author of *Patriotic Gore: Studies in the Literature of the American Civil War* (1962).

Chapter Four

Yes, They Were Traitors: John Logan's Great Conspiracy

I *t might seem easy to dismiss Gurowski as a crank whose prescience was sheer luck, but it's not so easy once you've read the speeches of proto-Confederate Congressmen who dominated the pre-Civil War United States. John Logan, an unjustly forgotten Union man who served in Congress with Jefferson Davis and his co-conspirators, compiled a permanent record of their statements and plots in the years leading up to the war.*

"The trouble with many of our generals in the beginning was that they did not believe in the war... They had views about slavery, protecting rebel property, State rights—political views that interfered with their judgments" *U.S. Grant speaking to John Russell Young.*[1]

What's missing from most recent histories of the Civil War is the radicalism of the Victorians. Back before the Lost Cause Myth won the storytelling contest and convinced America the war was just a tragic misunderstanding, people who'd actually had to deal with the real Confederates knew better.

The war was not the result of misunderstanding, but a decades-long effort by the Planters to control the Federal government in the interests of their "peculiar institution," slavery. When the 1860 election showed them that they were no longer in control of that government, they made their own country, a government by, for, and of slaveowners. They did this according to a plan they had been cooking up for decades

The most thorough account of their machinations is *The Great Conspiracy* (1886) by John A. Logan.[2] Logan knew first-hand what the proto-Confederates in Congress were doing because he'd been one of their colleagues before the war. When they chose secession and war, he became a Union officer and did pretty well for one of the "political generals" (men appointed to high rank in Union or Confederate armies because they represented important constituencies rather than for their previous military expertise). It's a myth, by the way, that all "political generals" made bad commanders, but that's another story.

---

1. *Patriotic Gore: Studies in theLiterature of the American Civil War* by Edmund Wilson first published 1931.

2. *The Great Conspiracy:Its Origin and History* by John A. Logan (1886; published as an ebook 2006)

Logan was from Illinois, a young lawyer. That was a very promising demographic if you were looking for Union officers in the western theater: ambitious young lawyers from Ohio or Illinois. Lots of them made the jump from oratorical fighting against the Planters in Congress to blasting away on the battlefields.

Logan's transition was especially dramatic. While serving in Congress, he volunteered as a private and fought at Bull Run, then resigned from Congress to command a regiment (31st Illinois). His war record was good, a lot better than that of many professional soldiers like Buell or McClellan, and he was on the field at a lot of Union victories: Fort Donelson, Corinth 2, Vicksburg, and the Atlanta campaign.

He started as a "Douglas Democrat," which meant roughly pro-slavery but pro-Union. It's hard to say whether those unsavory opinions were his own, or just the platform he needed, as an ambitious politician trying to get elected in Southern Illinois. Southern Illinois whites felt closer to Missouri and Kentucky than Chicago.

But Logan learned and changed. His experience of seditious Planter oratory in Congress shocked him, and his war service hardened him. Like Benjamin Butler, Stanton, and to some extent Grant himself, Logan moved toward radical Republican views: a hard war, the abolition not simply of outright slavery but of "every disability" imposed on African Americans, and resistance to letting the Planter elite retake control in the South.

His experience of proto-Confederate scheming began in 1858, when he first went to Washington. For two years, he listened to their speeches. Like many Douglas Democrats from the west, he was shocked to hear how arrogant and seditious the Planter aristocracy really was.

The Planters had been used to running DC for decades, and they saw no need to be coy. When they saw US demographics turning against their hegemony, they decided to provoke a war while they still had a monopoly on Federal offices. So their oratory just before the war—when Logan encountered them—got even more shrill.

Logan quotes Planter senators and congressmen at length to prove his thesis that secession was always the plan. Some of the quotes are amazing. You will never again believe the milquetoast lie that the war "was a tragic misunderstanding" after reading the words that Planter maniacs roared in those supposedly sacred halls of Congress. Here's a gem from Senator Louis Wigfall:

> "We may as well talk plainly about this matter. This is probably the last time I shall have the opportunity of addressing you. An invading army... cannot burn up plantations. You can pull down fences, but the negroes will put them up the next morning... Now I have told you what an invading [Northern] army cannot do. Suppose I reverse the picture and tell you what [a Southern Army] can do. An invading army in an enemy's country, where there is a dense popu lation, can subsist itself at very little cost; it does not always pay for what it gets. An invading army can burn down towns; an invading army can burn down man-ufactories; and it can starve [industrial] operatives.... You may bankrupt every man south of North Caroli-na...but the next autumn those cotton states will have just as much money and as much credit as they had before...Every time that a negro touches a cotton pod with his hand, he pulls a piece of silver out of it... "
> *(Chapter VII)*

Violence was the only form of discussion Wigfall's society really valued. It was his answer to political disagreement as well as to any imagined insult to his honor: "In a five-month period, Wigfall managed to get into a fist fight, two duels, three near-duels, and was charged, but not indicted, for killing a man."

Wigfall and his fellow slave owners in Congress were like 9th-century barons: drunk, arrogant, irrational, cruel, and extremely fond of violence. Preston Brooks became a Planter hero for beating abolitionist Senator Charles Sumner almost to death on the floor of the Capitol in 1856.

Wigfall is the voice of the stupider faction of the Planters, as shown in his dismissal of a naval blockade, one of those pre-Twitter Tweets that don't age very well: "I speak not of the absurdity of the position that you can blockade our ports, admitting at the same time that we are in the Union... You cannot use a blockade against your own people..." *(Chapter VII)* Well, Louis, see, if you *secede*, then you're not technically... oh what's the use. Not all the Planters were as stupid as Wigfall, but they were as arrogant and infatuated with violence as the answer to all dissension.

Jefferson Davis, leader of the Planter oligarchy, was not a stupid man. A profoundly evil and very strange one, but not stupid. But he seriously imagined that Southern armies would make a wasteland of Manhattan. You can hear the envy and the hate in his prediction, made on the floor of Congress, about how the war would go:

> "The grass will grow in the Northern cities, where the pavements have been worn off by the tread of commerce. We will carry war where it is easy to advance—where food for the sword and torch await

our armies in the densely populated cities..." *(Chapter VII)*

Davis waxes downright lyrical about "food for the sword and torch"—meaning "pillage and arson"—in northern cities. The next time a Lost Causer whines about how mean and unfair Sherman was in 1864, quote that passage.

Logan, the rookie congressman from Illinois, heard enough of this sort of ranting to be slapped out of his Democratic compromise delusions. Gurowski says over and over that it was the people of the north who forced their feeble representatives to confront, at last, the slaveowners:

> "[Fort] Sumpter [sic] bombarded; Virginia, under the nose of the administration, secedes, and the leaders did not see anything; flirted with Virginia... the people are taken unawares; but no wonder; the people saw the Cabinet, the President, and the military in complacent security. These watchmen did nothing to give a signal of alarm, so the people, confiding in them, went about its daily occupation. But it will rise as one man and with terrible wrath." *(April 1861)*

That was the story of the war: decades of oligarchical betrayal, then the shock of secession and a popular reaction too powerful to be diverted and neutralized. If the elite could have groveled to the slaveowners yet again in 1860, they would have. But the Planters were no longer interested in simply dominating. They had activated their long-standing Plan B: Secession.

As Gurowski says, "All the coquetting with Virginia, all the pre-
sumed influence of General [Winfield] Scott, ended in Virginia's secession, and in the seizure of Norfolk [Armory, where the Confederacy was able to confiscate 2,000 heavy cannon a few miles from the
Federal capital, without any Federal resistance]." *(April 1861)*

As a foreign radical, Gurowski saw the American landscape more
coldly and clearly than those who had gotten used to submitting to
the Planters. Logan, who began as a middle-of-the-road Democrat,
had clarity slapped into him by his time in Congress and four years
of war. *The Great Conspiracy* is his attempt to go back and document
the decades of secessionist plotting.

Logan's point is simple: the Planters never identified with the
United States except as a way of protecting their beloved "institution"
of slavery. Their eagerness to betray the US goes back, he argues, to
the 18th century. Logan's method is simple: quote these people and
let their own words damn them.

It's not an easy book to read because Victorians had a tremendous
appetite for legislative oratory. Logan wants to be thorough and my
God is he thorough. Every speech is quoted in full and some of them
consist of sentences 500 words long. As for paragraphs, they'd count
as novellas by most publishers' standards. Some of it is good oratory,
but reading this endless cascade of Ciceronian prose is exhausting. I'll
just try to boil down some of the key moments in the development of
this Planter Conspiracy, as documented by Logan.

Logan starts with the Constitution, but to make it manageable, I'll
just summarize his position:

> The fact of the matter is, that the convention that
> framed our Constitution lacked the courage of its
> convictions, and was 'bulldozed' by the few extreme

Southern slave-holding states—South Carolina and
Georgia especially.... A great majority of its delegates
were against not only the spread but also the very
existence of slavery... but surrendered as they always
did to the Planters.

The first big moment in Logan's chronology of nineteenth-century
treason is the Nullification Crisis. Nullification was the Planters' first
serious attempt to start a civil war. It began in the late 1820s and
climaxed a few years later. As usual, the epicenter was South Carolina; Senator Hayne of South Carolina made the case for Secession in
Congress in 1830. Haynes' view was that the United States was simply
a compact between the states, to be broken at will by any one of them.

In 1832, as Logan says, "[T]he leading men of South Carolina
unanimously passed an Ordinance of Nullification", which made
Federal law "null, void and no law nor binding on this state, its officers
or citizens." SC was quite ready to go to war over the matter, "[f]ur-
thermore, in the event of the Federal Government trying to enforce
[its laws], the people of South Carolina would thenceforth consider
themselves out of the Union." *(Chapter II)*

Hayne resigned from Congress and was elected Governor of South
Carolina, which made him, in his terms, a Head of State: "No allegiance [is] as paramount to that which the citizens of South Carolina
owe to the state of their birth or their adoption.... If the sacred soil
of Carolina should be polluted by the footsteps of an invader, or be
stained with the blood of her citizens, shed in her defense..." *(Chapter
II)*—then, of course, it would be time for what Hayne called "violent
remedy", or civil war.

Hayne called for an SC volunteer army to resist the US "invasion."
This would have been the civil war that the Planters wanted but, as

Logan explains, "...[t]here happened to be in the Presidential chair one of her [SC's] own sons, General [Andrew] Jackson."

Jackson was not someone you wanted to provoke. Like Colonel Kurtz, he was not a good man, he was not a kind man, but he was also not someone you wanted to annoy. Unlike the hopelessly feeble James Buchanan when facing the 1860 secession crisis, Jackson wasn't really interested in finding reasons to do nothing. He didn't ask his constitutional lawyers if he was allowed to come down hard on Hayne and his traitors. He just made arrangements, sending Winfield Scott to organize defense of federal installations in the state and described Nullification as "disunion by armed force," which, he clarified, "... is treason."

SC backed down because they knew Jackson all too well. A little anecdote to give you a sense of how Jackson worked: He was rail-thin and, when indulging in a duel (he was from SC, after all), he would wear a black overcoat and stand sideways so that his opponent would see nothing except a gun-barrel at shoulder height, a falcon-like face zeroing in on him and a black line extending to the ground. Jackson was a born killer, and you took his threats seriously or paid the price. So Hayne and the rest of the plotters backed down.

Jackson knew that the trade and tax issues involved in the Nullification Crisis had been only a pretext. As Logan says, Jackson "prophesied that the next [crisis] will be 'the slavery or negro question.' Jackson's forecast was correct." *(Chapter II)*

Jackson's plan to settle the matter by letting a few SC and Georgia Planters swing in the wind, *pour encourager les autres*, was let down by Congress, which passed a compromise bill caving to SC's pretexts. Just as Jackson had predicted, the Planters' next provocation was a pre-emptive defense of slavery, the only issue that really mattered to them.

Having decided to break with the Union when Lincoln was elected, their next move was to look for foreign support for their intended break. As Logan documents, the Southern faction in Congress began pursuing this strategy twenty years before the outbreak of the Civil War. The Planters' leader, John Calhoun, aided by James Murray Mason, began negotiating with Britain as early as 1841 to unite with a secessionist U.S. South to crush the North.

J.M. Mason, Calhoun's catspaw in this little-known scheme, was from one of the most privileged of all Planter families, the Masons of Virginia. Here's a list of some of his relatives:

> "He was a grandson of George Mason (1725–1792); nephew of George Mason V (1753–1796); grand-nephew of Thomson Mason (1733–1785); first cousin once removed of Stevens Thomson Mason (1760–1803) and John Thomson Mason (1765–1824); son of John Mason (1766–1849) and Anna Maria Murray Mason (1776–1857); first cousin of Thomson Francis Mason (1785–1838), George Mason VI (1786–1834), and Richard Barnes Mason (1797–1850); second cousin of Armistead Thomson Mason (1787–1819), John Thomson Mason (1787–1850), and John Thomson Mason, Jr. (1815–1873); second cousin once removed of Stevens Thomson Mason (1811–1843); and first cousin thrice removed of Charles O'Conor Goolrick. A first cousin

5 times removed was Betty Mason (1836-1899) first
wife of Gen. Edward Porter Alexander."[3]

This distinguished genealogy was Mason's only qualification. As
an individual, he seems to have been one of the most repellent creatures in the Planter bestiary. Carl Schurz, a brilliant German 1848 exile
who took a radical stand in the US, sketched Mason as "a sluggish
intellect spurred into activity by an overweening self-conceit". Many
other Planter frontmen were personally charming or at least enter-
taining. Mason was the exception. Nobody liked him.

One of Logan's biggest scoops is that Mason negotiated directly
with the British military, twenty years before the Civil War broke out,
to destroy the US. Most people know that Mason tried to reach Britain
in 1861, representing the Confederacy in what became known as the
Trent Affair. His ship, the Trent, was intercepted by the US Navy. He
and the other envoy John Slidell were taken off the ship and placed in
US custody.

Everyone agrees on that story. What Logan demonstrates is that
John Calhoun, the proto-Confederate leader, had employed Mason
in a similar role—conspiring to get British support for an attack from
Canada in unison with a Planter insurgency in the South—twenty
years earlier, in 1841:

> "Calhoun was indirectly, through himself (Mason), in
> secret communication with the British Government
> as far back as 1841, with a view to securing its pow-

---

3. Wikipedia entry for James M. Mason https://en.wikipedia.org/
wiki/James_M._Mason(9/19/2025)

erful aid in his aforesaid unalterable resolve to Secede from the Union..." *(Chapter III)*

Logan's proof is Mason's own account of this plot, which Mason blurted out once the South had finally seceded:

> "In the year 1841 the late Sir William Napier sent in two plans for subduing the Union, to the War Office, in the first of which the South was to be treated as an enemy, the second as a friend and ally. I was much consulted by him as to the second plan....[which] would, in eighteen or twenty months, have reduced the North to a much more impotent condition than it exhibits at present [1861]." *(Chapter III)*

I've done a lot of looking for what Mason was doing in 1841 and found a really interesting void. There's nothing! Until 1839 Mason was a congressman from Virginia, and after 1847 his life is well documented; but there's nothing about what he was doing in the early 1840s. As a book-length biography of Mason puts it, "[A]fter the 4th of March, 1839, he returned to his home in Virginia and to the practice of his profession. Several years then rolled by without making any marked change in his mode of life, or in the pursuits and pleasures that occupied and interested him."[4]

---

4. *The Public Life and Diplomatic Correspondence of James M. Mason With Some Personal History by His Daughter* by Virginia Mason. (Neale Publishing Company, 1906) p.48.

It ain't natural for someone as eminent as Mason, someone whose entire life is otherwise well documented, to have that big a gap in their CV during what should have been his best years.

As American demographics started to look bad for the Planter Elite, with immigrants from famine-stricken Ireland streaming into the northern cities and German peasants spreading out in what was then the north-west (Upper Midwest), the Planters began to see the US government as something they could no longer control and wanted no part of.

It seems odd, looking back, that slaveowners who had no interest in creating employment for non-slaves should have been angry or surprised that immigrants preferred northern farms or cities where they stood a chance of getting a job, but of course they were. You run into this kind of irrationality pretty often in the Planter Elite. As always when a group's economic interest conflicts hopelessly with its attempt to make sense of the world.

For example, James Murray Mason, "Least Lovable Man Ever Born" spent his career defending slavery for African Americans and also believed they were the "great curse of the country"—to the point that he insisted on bringing white Canadian servants to his house after the Confederacy lost.

Secession became, in the later 1840s, a mainstream middlebrow commonplace in the white South. DeBow's Review (first published in 1846) defended slavery in response to abolitionism. Their stance got more aggressive through the 1850s. As Logan puts it:

> "A Georgia delegate (Gaulden)... said [at a Democratic Partyconvention], 'I tell you fellow Democrats that the African Slave Trader is the true Union man (cheers and laughter.) I come from the first Con-

gressional District of Georgia. I represent the African
Slave Trade interest of that Section.... I advocate the
repeal of the laws prohibiting the African Slave Trade,
because I believe it to be the true Union movement."
(Chapter V)

Secessionists were intentionally making it impossible to compromise, even for someone like James Buchanan, who desperately wanted
to forward their interests. Jean Baker's biography[5] of Buchanan shows
that he was always an active agent for the Planter Elite but never
quite realized that they didn't want simply to dominate, they wanted
outright rule—and if that rule couldn't be over the US, it would be
over its dead body.

There was one safety valve still available to the weak northern
politicians desperate to placate the South: war of conquest against the
brown, Papist, Spanish-speaking people to the south. This is why the
Mexican War (1846-48) features so strongly in Civil War memoirs as
a lost Paradise where young American officers, the future officer corps
of Union and Confederate armies, could work together and earn their
medals against a weak, thinly populated Mexico.

This was the bargain the Planter Elite kept offering without ever
meaning to fulfill: "Give us more new slave territory to the south and
we'll stay in your damned Union."

After 1848, they had hundreds of thousands of square miles of new
territory to exploit. The question was, how much of it would be slave
states? The wrangling over that resulted in the Missouri Compromise

5. *James Buchanan: The American Presidents Series: The 15th President, 1857-1861* by Jean H. Baker(2004)

of 1850. This attempt to buy off the Planters failed like the rest, even though it was linked to the ultimate humiliation for northern anti-slavery voters, the Fugitive Slave Act.

This act, which required northern citizens to actively aid slave hunters even in downtown Boston, was meant to humiliate the New England abolitionists. Like all the Planters' bargains before 1861, it was designed to be confrontational. The proto-Confederates who dominated Congress before the war had no sense that their northern counterparts could ever be dangerous.

The Planter Elite wanted a final showdown. As Logan says, they were angrier at the attempted compromises of northern Democrats like Douglas than the outright abolitionism (as they imagined) of Lincoln:

> "In the South, the Democracy was almost a unit in opposition to Douglas, holding, as they did, that 'Douglas Free-Soilism' was 'far more dangerous to the South than the election of Lincoln; because it seeks to create a Free-Soil Party there; while, if Lincoln triumphs, the result cannot fail to be a South united in her own defense'.... The [Democratic Party was] split in twain, three-fourths of them upholding Douglas, and the balance, powerful beyond their numbers in the possession of Federal Offices, bitterly hostile to him, and anxious to beat him, even at the expense of securing the election of Lincoln." *(Chapter V)*

So Lincoln's election in 1860 was an occasion for hand-rubbing and conspiratorial chuckles among the Southern elite. Governor Gist of South Carolina rejoiced and started making preparations for war before Lincoln's election. Before the election results were known, he used the language of regretful necessity to hide his glee, "I am constrained to say that the only alternative left, in my judgment,is the secession of South Carolina from the federal Union."

Unlike the paralyzed federal government under Buchanan, South Carolina had no constitutional qualms:

> "He recommended the thorough reorganization of the Militia; the arming of every man in the State between the ages of eighteen and forty-five; and the immediate enrollment of ten thousand volunteers officered by themselves; and concluded with a confident 'appeal to the Disposer of all human events,' in whose keeping the 'Cause' was to be entrusted." *(Chapter V)*

Meanwhile, President Buchanan, went into long pettifogging orations about why it was simply not possible for him, given constitutional constraints, to do anything with the roughly 4,000 troops available, let alone (God forbid) raise any new troops.

I really don't know if there's ever been a case like that in all of world history. And, by the way, I can't stress enough that Buchanan was not "weak," as he's often portrayed with barely hidden homophobia. As Baker's biography makes clear, he was the most dynamic of the pre-war presidents, but his dynamism was entirely at the service of the Planter elite.

As is widely known, the Planters made a wee miscalculation and suffered a check but, as Logan wisely says in the last words of his massive tome, 1865 was only a setback for them, not a final defeat.

Writing in the 1880s, as the Lost Cause propaganda began to permeate American discourse, Logan saw that:

> "[a]lready the spirit of the former [Confederate] aggressiveness is defiantly bestirring itself. The old chieftains intend to take no more chances. They feel that their Great Conspiracy is now assured of success inside the Union. They hesitate not to declare that the power once held by them, and temporarily lost, is regained. Like the Old Man of the Sea, they are now on top and they: MEAN TO KEEP THERE–IF THEY CAN." *(Chapter XXXIII)*

Chapter Five

McClellan: "Top of His Class at The Point"

*W*hen *John Logan described the pervasive influence of the Southern planter elite in the U.S. government, much of what he said applied to the U.S. military, particularly graduates at the military academy at West Point, and above all to George McClellan, who was entrusted with the Union's largest force, The Army of the Potomac.*

"And fortune sings a mournful tune/ For those whose campaigns peak too soon"—*Bored of the Rings*[1]

---

1. *Bored of the Rings: A Parody of J.R.R. Tolkien's Lord of the Rings* by Harvard Lampoon, Henry N. Beard, Douglas C. Kenney (Signet, 1969)

Every Civil War fan knows the McClellan legend: the Little Napoleon came to West Point, the US military academy, at 15 and graduated "at the top of his class." It's a resonant story for everybody who was a striving American student, because of the way the rest of Little Mac's life turned out. As in, not so good.

There have been many attempts to rehabilitate McClellan's performance in the Civil War, but they're unconvincing. It's not every day that someone whose fans unironically compared him to Bonaparte is relieved of command, reduced to a mere presidential candidate (a losing one), and ends up sailing for Europe in a years-long sulk. That was Little Mac's tragic Third Act.

But was it really a big surprise? What did it mean, graduating top of one's class at The Point? It's worth looking harder at West Point in the mid-19th century, getting a sense of what that strange school drilled into its graduates.

For starters, McClellan was *not* first in his class. He graduated in 1846, just in time to join the land-grab invasion of Mexico, in a class of 59 students. The average graduating class was more like 40, since many of the boys who showed up at The Point couldn't take the "discipline," the intentionally petty, cruel regimen inflicted on first-years. It was meant to drive out the weak and the weird, a tradition that endured (or rather inflicted) more than a century after McClellan graduated.

"Discipline" at West Point in the mid-1800s meant a ten-page list of rules that could be enforced, changed, or ignored as the "upperclass-men" chose. Rules about everything: the uniform, the arrangement of one's belongings in the barracks, meals, salutes, bedtime. All the petty sadism you find in boot camps everywhere.

But West Point was not like other training camps. It wasn't meant to produce mere enlisted men, but officers in the United States Army. That meant something in 19th-century America; there were no scholarships, no college football teams, no special entry for the poor-but-deserving. It was an elite boarding school, but that didn't mean the discipline was easier. Cadets were not beaten, as enlisted men often were, but they were put through the same petty ritual humiliations that are now reserved for rich people's kids who get in trouble and are sent to some off-grid punishment camp to be abused without publicity.

Your rank in the graduating class at The Point was a public stat, so everyone in the military community knew their numbers. The way people who peaked at the SAT used to make sure you knew their score after a few minutes of conversation.

In other words, McClellan, in the consensus view, had a great resume. What did that mean, in mid-19th century America? Not much, in terms of predicting which graduates would be good commanders in the Civil War.

U.S. Grant, the biggest success story of the war, was a decent student at West Point (not the complete failure he's often considered), graduating 21st out of the 39 graduates in the class of 1843. He was the bulge of the West Point bell curve, a decent cadet but nothing special.

McClellan, who graduated three years after Grant, did seem special. Although he was not the top student of his graduating class, he graduated in second place, which was considered a big deal. And McClellan didn't lose out for first place because of the obvious defects he showed in combat. West Point in the 1840s devoted only nine class hours to military affairs:

> "In only nine hours of class time [cadets] would learn
> all that West Point intended to teach them about army
> organization, order of battle, laying out a military

camp, reconnaissance, outpost duties, attack and defense, and the principles of strategy."[2]

Nine hours, out of the thousands of hours that cadets confined
to the grounds for four consecutive years devoted to their studies!
Once you know that, it doesn't seem so surprising that so many West
Pointers did so badly at real war, and that so many mere civilians put
on the uniform, learned on the job, and surpassed them.

West Point didn't test you for real-time command stress, like the
Kobayashi Maru exam in Star Fleet Academy. If it had used a simple
low-tech stress test, using 19th-century technology, like putting Little
Mac in a room and having some big drill sergeant yell at him for five
minutes, he'd have broken and been instantly expelled, if not institu-
tionalized.

The Academy tested other sorts of stress, like the ability to obey a
multitude of petty rules about neatness. And, of course, Math. Neatness and math—I would have lasted roughly five seconds there. West
Point taught applied, not pure math. The Point was an engineering
school. Those graduates who showed promise as engineers were the
elite:

> "...Class standings [at West Point] were so impor-
> tant [because] they dictated where [the graduates]
> could and could not go and would likely set the

---

2. *The Class of 1846: From West Point to Appomatox- Stonewall
Jackson, George McClellan and Their Brothers* by John C. Waugh
(Random House Publishing Group, 1999). "Gone Are the Days
of Our Youth."

compass for...army careers. Most of them would have preferred the engineers, for prestige's sake. But few were called, for engineers were 'species of gods.' So rarefied was that branch of the service that even top-ranked cadets in some classes were not recommended for it after graduation. But [McClellan's class of 1846] had been outstanding. The [top four graduates] had been recommended [as engineers]...Just below [them]...stood...topographical engineers. The 'topogs' were but 'demigods' in the order of things, but they were god enough." *(Class of 1846, "Gone Are the Days of Our Youth.")*

This is very surprising for anybody who tried to be a little eco-warrior in the 1970s, because the US Army Corps of Engineers at that point had degenerated into a sleazy conduit for pork-barrel appropriations and for anybody who could use a slide-rule and felt, on looking at a pristine salmon stream, "What this needs is to be concreted and channelized." It was most definitely not an elite branch of the service.

Nor would engineers have been the cream of the crop at most other 19th-century military academies. The more dashing branches of the service, like the cavalry, usually drew the top of the class.

But at West Point, some strange-seeming skills useful only to an engineer were important. For example, McClellan had to settle for second place in his graduating class because he couldn't draw very well. The top-ranked cadet in his class was Charles Seaforth Stewart, who did nothing worth remembering in the USCW. He was, however, very good at drawing. (If you're looking for questions for a trivia quiz, that would be a good one: "Who beat out George McClellan for top of the class, West Point class of 1846?").

The most interesting thing about Stewart is that he gets no attention. Not then, not now. Odd, when he topped McClellan at West Point. John Waugh's book *The Class of 1846* mentions Stewart only 17 times, almost always in comparison with the "real star" of the class, McClellan, who gets 825 mentions:

> "There was no doubt in the class's mind who their real star had been. It had not been Stewart, despite the final rankings. It had been the born and bred little gentleman from Philadelphia [George McClellan]." (*US Military Academy, Annual Reunion, 1904*)

Stewart may have been a quiet obsessive, without the gigantic ego you needed to stand out in that crowd. It's difficult to say, because there's very little detail about him anywhere. One hint is provided by an alumnus when commemorating his life:

> "It was peculiarly characteristic of Captain Stewart that he would never delegate to another what he could possibly perform himself, and he was indefatigable in all his official work.... It was this attention to detail, and unnecessary attention at times, and this unsparing although unassuming energy, that consumed his power and limited his ultimate service." (*US Military Academy, Annual Reunion, 1904*)

When you translate those guarded phrases into contemporary pathologizing, colloquial English, you get something like "obsessive, nervous, withdrawn." The one hint of a flaw, "...unnecessary attention at times…" is very loud in context. We would probably call Stewart autistic or "on the spectrum" or whatever the nomenclature is allowing people to say now. It tells you that Stewart wasted his career on details, which explains why he spent the Civil War years designing fortifications for the Union against a highly unlikely, not to say impossible, Confederate naval attack.

Stewart, it seems, was just too quiet to interest The Point or its chroniclers. The atmosphere up there was something like two-thirds ego, one-third testosterone, with a little nitrogen to keep the whole thing from spontaneously combusting.

McClellan's class included Southern wannabe cavaliers like Dabney Maury, Jeb Stuart, and George Pickett, who considered themselves the descendants of the losers in the English Civil War, destined to fight (and lose again) to the Puritans of New England. It was self-fulfilling (or rather self-inflicted and self-serving) nonsense, of course. But that counts too.

Along with the Southern cavaliers, there were the smart, provincial kids like Grant, who wanted a chance to jump an entire social class in one generation by becoming an officer and a gentleman.

And then there was the most dangerous clique of all, northern-born cadets like McClellan who adopted the Southern Cavalier ethos and conducted themselves like "born and bred" gentlemen, writing home sneering letters about farmers, African-Americans, and anyone else their cohort deemed inferior.

It was not easy to stand out as an arrogant snob in a group of adolescent males, virtually all of whom would be prescribed Adderall and scheduled for weekly appointments with a mental-health professional if they enrolled today, but McClellan managed it.

West Point was, after all, a finishing school, which is standard for military academies. The objections made to "elite" colleges in the US

now are essentially grumbles that their students are being taught a
new etiquette that is alien and obnoxious to an older generation. The
Point differed only in that it wasn't intended so much to enforce the
appearance of sensitivity as to inculcate a more genteel manner, under
which rested an absolute sense of what would now be called privilege.

McClellan, who showed up at The Point aged fifteen, the darling of
his family, was the perfect template: good at math, but no mere swot
like Stewart; a perfect little "gentleman," full of petty snobbery in a
way that warmed the hearts of his southern "Cavalier" schoolmates.
He soon became a favorite, admitting in a letter home that "...the
manners, feelings, & opinions of the Southerners [at West Point] are
far, far preferable to those of the majority of the northerners at this
place..."

McClellan was a perfect recruit for this clique: easily flattered,
constantly in need of reassurance, with a fragile sense of superiority
floating over an infinite terror (as he showed in his letters during the
war, and all too clearly in his strategic decisions). As such, he was an
easy catch for Planter sons like Pickett and Maury, who became his
best friends at the Academy.

There were several ways of construing "gentleman," a key term.
The cavaliers had no monopoly on it. Most of the northern cadets
considered themselves "gentlemen" as well, though in a different, less
gaudy sense.

Certainly Charles Seaforth Stewart, born on a USN ship captained
by his father, descended from many generations of Ulster Presbyteri-
ans who dominated their part of upstate New York, would have grown
up with the conviction that he was a gentleman. But his definition
would not have satisfied the Planters' sons like Maury or Pickett, who
believed that "...to fight like a gentleman, a man must eat and drink
like a gentleman...".

The New England definition of "gentleman" involved less drinking, more self-control and piety, reaching an extreme in cadets like Stewart, who seems to have been so full of these attributes that he didn't even register on his classmates until he graduated at the top of the class.

McClellan took the Southern concept of "gentleman" very seriously, with its touchy emphasis on personal honor. It suited his fragile, swollen ego. He was younger than the average cadet, and young for his age as well: gregarious, eager to join in group activities, a born snob and joiner. Southern cadets encouraged his snobbery toward their rough classmates from "The West" (Ohio, Indiana, Iowa.) McClellan was also fussily neat about his clothes, prickly about his personal honor, vain about his lineage—in short, the perfect "little gentleman."

And he was an excellent engineer, though never quite as good as Stewart. In letters home, McClellan alluded to Stewart in the petulant, anxious tone which was to become all too familiar in the war years: "I do not get marked as well for as good (or a better) recitation as the man above me..." The stress of competing against Stewart was already revealing the essential McClellan.

McClellan followed his Cavalier buddies' lead, even in love. In a rather creepy, ingrown love triangle story, he courted Mary Ellen Marcy, who was engaged to McClellan's classmate and friend, future Confederate general A.P. Hill. She clearly preferred Hill, but McClellan was her father's choice because he was planning to leave the army and go into the railroad business. That meant big money. Staying in the army, as Hill intended to do, meant living frugally, with endless transfers around the more remote parts of the continent. So after seven years of being screamed at by her father and courted with grim determination by Little Mac, she surrendered and married McClellan in 1860.

After McClellan graduated in 1846, he built an impressive CV. After serving in the Mexican War from 1846-48 (without much distinction), he took part in many army-funded surveying and construction projects in the West. Then, in classic manner, in 1857 he quit the army and jumped to the top echelon of the railroad business, the IT business of the mid-nineteenth century.

At heart, McClellan remained the callow West Point snob he had always been. His wife had been the center of a clique of male admirers, most of whom sided with the South when war came. He never changed his youthful, snobbish opinions about everyone his Southern 'cavalier' buddies taught him to despise: "Nothern mud-sills," urban workers, and the enslaved.

When war came, McClellan was put in command of Union armies in the Kanawha Valley, at that time still part of Virginia though at the end of the war it became part of the new state West Virginia. Readers of John Beatty's memoir, *The Citizen-Soldier* will recall that Beatty served in the Kanawha campaign under McClellan and was massively unimpressed. He considered the real hero of that campaign to be McC's subordinate William Rosecrans. McClellan did his best to get Rosecrans's little force wiped out at Rich Mountain with his trademark move: refusing to advance to support his subordinates.

McClellan's battle paralysis at Rich Mountain is described in Beatty's memoir. The plan was for Rosecrans to lead four regiments along a trail and then pop out behind the Confederate fortifications. According to the plan, McClellan's larger force in front of the Confederates would advance as soon as they heard "shots in the rear of the [Confederate] fortifications."

Rosecrans, following a local Unionist guide, does his part perfectly, appearing suddenly in the Confederates' rear. Beatty and his regiment are waiting with McClellan. They hear the shots, followed by bigger volleys of gunfire and artillery. But McClellan will not order an advance. Beatty describes what happened:

> "Every man sprang to his feet, assured that the moment for making the attack had arrived. General McClellan and staff came galloping up, and a thousand faces turned to hear the order to advance; but no order was given. The General halted a few paces from our line, and sat on his horse listening to the guns, apparently in doubt as to what to do; and as he sat there with indecision stamped on every line of his countenance, the battle grew fiercer in the enemy's rear...[We thought] if the enemy is too strong for us to attack, what must be the fate of Rosecrans' four regiments, cut off from us, and struggling against such odds?"
> *(July, 1861)*

In this account of McClellan in command before he was given the Army of the Potomac, you can see the outcome of Antietam and the Seven Days battles, with Fitz John Porter whispering in his ear like Lucifer, convincing him not to send in his reserves and destroy Lee's forces.

McClellan should not have been allowed to stay in command after Rich Mountain. But after leaving Rosecrans in danger, he actually became the hero of the battle because he realized that a gift for hype was more important than winning battles.

As a promising, pampered graduate of The Point, McClellan had been selected as an official observer of the Crimean War in the mid-1850s. He'd seen what the telegraph could do—not to win battles, but to make a commander's reputation. And when it came to public relations tech, McClellan truly was the second coming of Napoleon.

So as soon as the Battle of Rich Mountain was won by Rose-crans, McClellan, who'd played no part in that attack, telegraphed to Washington D. C., "We have annihilated the enemy in Western Virginia...The troops defeated are the crack regiments of Eastern Virginia, aided by Georgians, Tennesseeans, and Carolinians. Our success is complete, and secession is killed in this country."

It worked. McClellan was a speed-of-light hero. The administration was "charmed with [McClellan's] activity, valor, and consequent successes." The New York Times, with Thomas Friedman-like discernment, declared itself "...very proud of our wise and brave young Major-General. There is a future before him...which he will make illustrious."

This all feels very familiar, far too familiar. The world of the blue-chip undergrad, the SAT hero, morphs easily into that of the far more lethal self-promoter, and glides toward the corpses lining the fields of Antietam.

Chapter Six

Poe at West Point

I *t's useful to have a point of comparison to McClellan's West Point experiences, and luckily we have one in the early life of a familiar figure from American literature, one E.A. Poe.*

You come across some amazing stories, researching the creatures who formed the officer corps of the US Civil War. I just ran into a wonderful collision of literary and military history. It involves Edgar Allen Poe, which almost promises some wild fun.

Poe's one of those American writers who were better appreciated in Europe than over here, like Philip K. Dick and Jack Vance. He has a lot in common with those two, in fact. But he also partook of the awe-inspiring, and sometimes just plain awful, demonic energy of his generation. Like a lot of the most bloodthirsty elite youth of the period, Poe ended up at West Point, the US Army Academy on the Hudson River north of NYC.

West Point started as one of Alexander Hamilton's pet projects. I guess that's a point in its favor—for people who saw the musical and got their history from it. If you know Hamilton's real story as a lifelong supporter of militarized oligarchy and his enthusiasm for using a professional army to destroy popular uprisings, his support for a military academy isn't a surprise, let alone a point in his favor.

He and the rest of the American elite were driven by the urge to be just like European powers, in all the wrong ways.

Hamilton was thinking ahead, as usual. As Secretary of the Treasury Hamilton bought the land for the academy, a point projecting over the Hudson, in 1790. Nice location, good site for a battery (which it was in the American Revolutionary War).

The man Hamilton bought it from was a fellow general. This was how Hamilton imagined good government, a very Cheney-esque vision of tax money passed eternally among a small military/oligarchic elite. Though, to be fair, Hamilton was a good soldier, unlike Cheney. In fact, you can pretty much assume that everyone in that earlier elite was physically brave. It's the rare cowards who stand out as the anomalies. In that way at least, times have changed; our militarists aren't even brave.

West Point was designed to create aristocratic Spartan warriors, and it generally did—in a few ways that were good, and a lot of others that turned out to be disastrous when put to the test in the Civil War.

Cadets had to be recommended by a member of Congress, meaning most of them were surplus sons of locally influential families, especially those in the slave states. There were a lot of such sons, since the average American woman produced seven children in 1810.

And since the offspring of wealthy families probably survived at a higher rate than those of the poor, a lot of well-born boys applied for the West Point of 1834, which should've been E. A. Poe's graduating class (he was born in 1809).

West Point seems to have operated pretty loosely for its first 15 years, but in 1817 a new commandant, Colonel Sylvanus Thayer, took over and clamped down hard. Thayer set the rules that Americans still think of when they imagine West Point: marching to meals, getting screamed at by senior class men, harsh punishments for minor infractions—a set of brutal initiation rites and grim, anti-intellectual rules.

Cadets were forbidden any books not directly related to their courses at the academy, which were engineering, math, horse-riding, and French. Why French? First because the American elite was in many ways more cosmopolitan or at least trans-Atlantic than it is today. Thayer himself studied for two years at the Ecole Polytechnique in Paris. Second, because the French Army was still the paradigm of military brilliance, and most military manuals were in French. And lastly because speaking French, like riding well, was one of the things that defined a gentleman, as in "an officer and a...".

You might suppose that all these grim Spartan rules would've crushed a poor, fragile bohemian like the young E. A. Poe. Hah! By the time Poe entered the Academy in 1830, he'd been through far worse than a few upperclassmen could dish out.

Poe's early life seems wild now, but then a lot of things that seem wild now were pretty standard in the precarious, mobile world of early 19th-century America. The child of a couple of minor actors who both died before he was three, he was taken in—not adopted, only taken in—by a friend of his mother's. His status was precarious but not fixed.

He did write poetry, but that could be forgiven because he was witty and athletic. When he heard Byron's boast about swimming the Hellespont, Poe checked the World Atlas, saw the modest distance involved, and spluttered "We used to swim ten times that far every day in Chesapeake Bay!"

But he had stepfather problems. His stepmom liked him fine, contrary to Indo-European narrative tradition, but his stepdad never forgave the kid for catching him fooling around with other women.

Stepdad took it out on the boy in classic 19th-century fashion, by
being especially tough but fair with him, emphasis on the "tough."

Poe started his undergraduate career at the University of Virginia,
where he gambled away his allowance. That was normal enough, but
the other students could afford it since they were mostly planters' sons
with deep, slave-funded pockets. Poe's allowance came from a stepdad
who was probably looking for a reason to cut him off. He did, and Poe
had to quit the university.

In 1827 he enlisted in the US Army, lying about his age and his
name (he enlisted as "Edgar A. Perry." Not the most cryptic of aliases).
"Enlisted" is the key word. As the son, or at least ward, of a wealthy,
respectable Virginia family, he wasn't supposed to become a mere
soldier.

Grant, in his memoirs, describes going home in uniform and encountering the general disdain for the profession, from a street kid:

> "Soon after the arrival of the [uniform] suit I donned
> it, and put off for Cincinnati on horseback. While I
> was riding along a street of that city, imagining that
> every one was looking at me, with a feeling akin to
> mine when I first saw General Scott, a little urchin,
> bareheaded, footed, with dirty and ragged pants held
> up by bare a single gallows—that's what suspenders
> were called then—and a shirt that had not seen a
> wash-tub for weeks, turned to me and cried: 'Soldier!
> will you work? No, sir—ee; I'll sell my shirt first!!'"[1]

---

1. *Personal Memoirs of U.S. Grant.* Chapter II

Enlisting was such a disgrace, such a blot on any boy from a rich family, that I wonder if Poe enlisted as a way to shame his stepdad—the ol' "Look what they made me do!" move beloved of offended adolescents.

But then a funny thing happened: Poe turned out to be a good soldier.

The US Army in that era wasn't exactly an elite institution at enlisted-man level. It was where you went if your other options were gone, or if you were a starving immigrant. Poe, the wellborn poet, must've stood out to his superiors as one of their own, somehow misplaced among the rabble.

He was promoted to Sergeant Major, the highest noncom rank, by 1829. He was placed in the artillery, traditionally the arm of choice for smart kids, and had pretty soft duty stationed at Fortress Monroe in Virginia (Monroe was also one of Robert E. Lee's first postings and Bobby Lee was the darling of the Virginia military establishment). Weird as it may seem, Poe seems to have liked the Army.

But, seeing commissioned officers treat him as an equal, he realized that his education, manners, and that powerful stepdad, all meant he too could become an army officer and be respectable enough to pursue a career in poetry.

There's a detailed history of Poe's military career by a guy who was killed in Iraq in 2006, *Private Perry and Mister Poe: The West Point Poems, 1831.*[2] It proves once and for all that Poe was a good soldier,

---

2. "Private Perry and Mister Poe," *The West Point Poems, Facsimile Edition*, by William F. Hecker III (Louisiana State University Press, 2005), pp. xvii-lxxv

enjoyed the army, and dreamed of military glory. Just don't read the parts about Poe's poetry.

But as a mere soldier, Poe was stuck. He had enlisted for a term of five years and once you'd signed enlistment papers you couldn't just quit unless you found a substitute willing to enlist in your place. So Poe got another soldier whose enlistment was up to reenlist in Poe's place—for a fee of $75.

The rest was easy. He groveled to his stepdad, got the old man to use his influence, and soon had an appointment to the academy. Poe passed the West Point entrance exam, which had tripped up many a pampered rich boy. In the Spring of 1830, he was enrolled in West Point, going through the usual ritual humiliations dished out to freshman cadets without being especially bothered.

If only he'd applied a year earlier, he could've met another famous graduate, one Robert E. Lee. Lee went through The Point without getting a single demerit. As Bad Ash would chant, "Goody little two shoes!"[3]

Poe's brief career at The Point wasn't as squeaky-clean as Lee's, but he was no drunken fuckup either. He was, for one thing, just plain smarter than everybody else. They crammed for the daily recitals in every subject; Poe just looked at his book for a couple of minutes and was set. He devoted the resulting free time to educating his fellow students...in the sense that we true educators understand it, to wit: explaining how wrong everybody else is:

> "Poe filled the study hours in barracks with vigorous
> discourse—generally a monologue—on English liter-

---

3. A character in Sam Raimi's *Army of Darkness* (1992)

ature, and in composing poems lampooning the officers at the academy. Both his memory and his knowledge of English literature were superb, and he could
hold forth on the English classics for hours.... The
whole bent of his mind at that time seemed to be
toward criticism—or, more properly speaking, cavi
ling....Whether it was Shakespeare or Byron, Addison
or Johnson—the acknowledged classic or the latest
poetaster—all came in alike for his critical censure....I
never heard him speak in terms of praise of any English
writer, living or dead."[4]

Then Poe got sick. No one knew what was wrong, but he looked so
old that the rumor went around that he was actually Poe Senior, and
had stolen the appointment given to his son. (Poe Sr. had in fact died
when E. A. Poe was three years old.)

Illness seems like the most plausible explanation for Poe's burnout
at West Point. I've had a couple of slow, tough-to-diagnose illnesses,
and God, do they age you and drain you of any enthusiasm for any
damn thing in the world. You can't do much but talk, and your talk
is bitter because something's wrong inside and nobody can help. It's a
lot like being lowered into a twilit ocean in a shark cage, sans breathing
gear. Well, that may be mere projection, but it makes more sense to me
than the notion that Poe was too emotionally frail or "poetic" to make
it in the rough crowd at The Point.

---

4. "Cadet Edgar A. Poe." by Tom Johnson. American Heritage.
   Volume 27, Issue 4. 1976.

As Hecker (the biographer who died in Iraq) shows pretty decisively, Poe was at least as tough as his classmates, about a zillion times
smarter, and with years of experience of military routine they lacked.
That "sensitive poet" stereotype is pretty shaky anyway. The great
early poets in this language were a downright murderous crew, and it's
a big mistake to conflate "glib" with "sensitive."

Yet these psychological explanations for Poe's problems at West
Point live on, just as the notion of psychosomatic illness lives on in
popular culture, though the conditions attributed to it keep turning
out to have bodily origins. It's an easy out for a dawn culture like ours,
centuries from understanding anything beyond germ theory.

People are still saying things like this:

> "The academic atmosphere at West Point was as se-
> vere as its military discipline; despite the excellence of
> its engineering education, the academy was hardly a
> sympathetic surrounding for a young romantic poet
> searching for his voice."[5]

This doesn't make sense. Poe did well as an artilleryman in the regulars, never showing up in the disciplinary lists. He did well in math-
ematics at The Point (17th in a class of 67). He was not "searching for
his voice"; that's some kind of twentieth-century back-formation. Poe,
like most non-Laker poets, found his voice at about the same time it
broke. He knew all about his damn voice before he started shaving.

It's true that he gave up on West Point halfway through his first year,
but there's just no evidence it was because he couldn't hack it there.

---

5. American Heritage, June 1976, Volume 27, Issue 4

Most likely, he was sick and getting sicker. And he was broke again, in another environment like UVa, where most of his classmates were rich brats.

And there was a fresh humiliation: the soldier who Poe had promised $75 to reenlist in his place claimed he'd never been paid, and wrote to Poe's stepdad demanding the money. The stepdad sent the money, then sent Poe the whole correspondence, with a note on the lines of "And now I disown thee! Never darken my door again, ungrateful child!"

The stepdad had also recently married again, the new wife producing an heir in a suspiciously few months. That meant that any hope of eventual reconciliation (and inclusion in stepdad's will) was gone, along with the dream of an officer's life.

Being an officer was nice—if you had family money to finance the endless card games and brothel visits which livened up garrison life. Without a private income, Poe must've seen ahead of him a lifetime of hanging out with guys who were ten times richer and 20 times dumber than he was, being endlessly humiliated by his poverty by people who'd never done a day's work in their lives.

So, there's Poe in the winter of 1830-31, in a grim unheated barracks at West Point: deathly sick, dead broke, humiliated and rejected by the only family he had, looking forward to a long career of further humiliation. All in all, it strikes me as plenty of reasons to want out, without resorting to any mid-20th-c. psychology or pop literary history about "sensitive" poets.

So, in January 1831, Poe started trying to get out. Once again, though, he was in an institution that didn't allow him to leave. Cadets could not resign. Poe, in a delightful move, invented his own Catch-22 solution: he stopped showing up for formations, classes, meals...he just plain stopped showing up.

Even then they didn't boot him, which suggests his prior rep at The Point must've been pretty good (it helped that most of the cadets were precociously alcoholic, which Poe was not). An upperclassman finally marched into his room and told him to shape up and go to Chapel. It's interesting they put their institutional foot down over chapel, rather than classes or formation. That's those proto-Victorians all over.

He refused. They yelled some more. He still refused, all those years of being yelled at by stepdad and army officers turned out to be useful for holding the line. He was court-martialed and dismissed from the academy.

He had one last request to make of the commandant, the legendary Sylvanus Thayer, the Lycurgus of the minor-league Sparta called West Point. All Poe asked was an introduction to Thayer's military contacts in France. (Remember, Thayer had studied at the Ecole Polytechnique for two years and had friends in the French Army.) Here's Poe's letter:

> "Sir, Having no longer any ties which can bind me to my native country—no prospects—nor any friends—I intend by the first opportunity to proceed to Paris with the view of obtaining thro' the interest of the Marquis de La Fayette, an appointment (if possible) in the Polish Army. In the event of the interference of France in behalf of Poland this may easily be effected—at all events it will be my only feasible plan of procedure. The object of this letter is respectfully to request that you will give me such assistance as may lie in your power in furtherance of my views. A certificate of 'standing' in my class is all that I have any right to expect. Anything farther—a letter to a friend in Paris—or to the Marquis—would be a kindness

which I should never forget. Most respectfully, Yr.
Obt. St., Most respectfully, Yr. Obt. St., Edgar A Poe"
6

That article I've mentioned describes this as a "strange" letter.
This is what's most interesting about the way historians of our era
try to deal with Poe, or at least did before Hecker's rehabilitation
of Poe's military career: as if Poe were babbling incomprehensibly,
typical sensitive poet/loon that he was. If you know what young men
across Europe and America—all the Americas, including the Spanish-,
French-, and Portuguese-speaking Americas—were thinking, arguing,
and dreaming about in early 1831, there is nothing at all strange about
the letter.

It's a little thing called the November Revolution in Poland. The
authors of that 1976 article in American Heritage magazine don't
seem to be familiar with it, but it was a sensation across the world in
Poe's day. There's no space here to go into the genuinely heroic, tragic
story of the Polish Rebellion of 1831. It was crushed by overwhelming
numbers of Tsarist troops, and the aftermath was brutal. That too
gave it luster. And, for the few months the Poles held out against
the Russian Empire, the Polish "Novembrists" stood as a last hope
to young, progressive Europe and America, especially young French
people who'd seen their 1830 insurrection crushed. The Marquis de la

---

6. **Edgar Allan Poe to** Colonel Sylvanus Thayer — **March
10, 1831 (LTR-030) https://www.eapoe.org/works/let-
ters/p3103100.htm**

Fayette—yes, the same guy who helped in the American Revolutionary War—was a leading man in the pro-Polish agitation in France.

So if you're Edgar Allan Poe, kicked out of West Point in the winter
of 1831, with military experience, a slow, wasting illness, a real love of
military glory, and an urgent need of a way to make a living, asking
your base commandant, a man with first-rate contacts in France, to
help you get connected in Paris with a view to joining the Polish
insurgency seems rational and noble.

That's why I want to quote the way this American Heritage article, written during America's bicentennial, deals with Poe's letter to
Thayer. It's worth reading this in detail, because it shows something I
see again and again:

> "Some of the scholars who know of the existence of
> this letter regard it as enlightening concerning Poe's
> unbalanced frame of mind at the start of an ill-under-
> stood portion of his life. Others, pointing to the pos-
> sible pun in 'Poe-lish,' consider it an obvious though
> somewhat ill-conceived joke. In history it is general-
> ly foolish to say that we will never know something,
> but this curious letter will probably remain a mystery.
> Colonel Thayer, at any rate, did not take it serious-
> ly. As though in concurrence with John Allan [Poe's
> stepfather], he never answered it."[7]

---

7. "Cadet Edgar A. Poe." by Tom Johnson. American Heritage.
   Volume 27, Issue 4. 1976

This, THIS, is why I couldn't hack my own West Point, grad study in the Humanities at Berkeley in the 1980s. This is the sort of provincialism which uses its own ignorance to create explanations which would not even be necessary if the authors had any knowledge of the period they claim to study.

In the 1980s, true, it was usually veiled in high-theory gibberish, but as this pre-theory, tweedy paragraph shows, the same provincial arrogance informed American Humanities' mainstream long before shake-n-bake intellectuals started quoting Foucault.

First, these unnamed "scholars" read the letter as proof of "Poe's unbalanced frame of mind." Why? He was interested in a military career, has discovered he literally can't afford one in his homeland, and so volunteers for a noble, popular cause abroad. That's absolutely standard behavior for educated young 19th-century men around the world; Byron fighting in the Greek War of Independence is probably the most famous example. Hell, Lafayette himself followed a similar course, which is why there's a town named after him in all 50 states.

Yet this article insists again, a few sentences later, that Poe's letter is "a mystery" and invokes the inscrutability of "history" to suggest we can never hope to understand Poe's request: "...this curious letter will probably remain a mystery."

Somehow, mainstream American consciousness became more provincial in the third quarter of the 20th century than it had been in Poe's time. I don't know how that's even possible, given the jump in media technology and ease of travel, but it happened. I see it over and over, reading histories of the Civil War written from the middle of the last century. In some way, American exceptionalism reached its peak back then, perhaps matching America's almost unique post-1945 prosperity. It seems to be waning now that the US is back in the middle of the pack. It's harder to ignore the rest of the world when you're just another piece of it. That's my guess anyway, but maybe I'll just settle for "...this curious mid/late 20th century stupidity will probably remain a mystery."

That's not even the worst of this accursed paragraph. I've saved that bit for last. I refer, of course, to "some scholars'" thesis on why Poe was drawn to...a certain east-European country. That's right: some tenured swine, and I wish I knew his name, came up with the brilliant notion that Poe was making a tedious academic pun. "Poe"...."Poe-l and"...geddit? Geddit?

What truly unbalanced, provincial pedantry and ignorance is re-vealed in that notion. I bet the bastard who came up with it got a million laughs over the box of Chardonnay in the English Lounge. I bet he thought Stevens was "dry" or "abstruse." I bet he considered *The Great Gatsby* a great novel. I bet...

But I digress. The good news is that Boomer stupidity peaked around 1976, when this article was published. America is poor-er, meaner, more thoroughly doomed now, sure. But as Hecker's book shows—as Hecker's whole life story shows the demographics of American stupidity have changed. It has moved up the pyramid, to the very top. But in compensation (of a sort), the middle ranks are not as stupid as they were.

Army Maj. William F. Hecker III, 37, of St. Louis, Missouri. Hecker died in An Najaf, Iraq, when an improvised explosive device detonated near his HMMWV during convoy operations. He was as-signed to the 3rd Battalion, 16th Field Artillery, 2nd Brigade Combat Team, 4th Infantry Division, Fort Hood, Texas. Died on January 5, 2006.

Chapter Seven

A Traitor Writes Home: McClellan's Harrison's Landing Letter

*W*hen it comes to McClellan-bashing essays, one is never enough. The Peninsula Campaign (March to July 1862) should have been McClellan's great hour. He had all the advantages. He could have taken Richmond easily, but he was, unfortunately, McClellan.

I've talked about Little Mac at West Point, but to get the full horribleness of the man, you need to see the letter he wrote to Lincoln in July 1862.

When he wrote this letter, McClellan was cowering in Harrison's Landing, backed up against the James River in Virginia, waiting to be annihilated by a massive Confederate army. That makes it sound like he was alone. In reality he led the biggest, best-equipped army ever assembled in North America. But McClellan was always alone, either boasting or cowering. And in the presence of an enemy, he cowered, no matter how many divisions he had around him.

Lincoln summed it up when he looked at McClellan's camp at Harrison's Landing. He asked an aide named Hatch, "Hatch, what is all of this?"

"Why, Mr. Lincoln, this is the Army of the Potomac."

"No, Hatch, no. This is General McClellan's bodyguard!"

McClellan's letter to Lincoln was written at the end of the Seven Days Battles, the last phase of the Peninsula Campaign, a grand plan to take Richmond, the Confederate capital, by using the US Navy. The whole Army of the Potomac, a hundred thousand men with all their equipment, boarded transports and steamed south along the coast. They entered the long estuaries of the York and James rivers, steamed upstream almost to Richmond, then landed and marched towards the city.

It was a needlessly elaborate, furtive way of getting to Richmond. A better plan would have been marching right at the Confederate army, overwhelming it, then marching into Richmond and "hang[ing] Jeff Davis from a sour apple tree," as suggested in the lyrics of "John Brown's Body."

But that wasn't McClellan's way. He could not imagine a straightforward solution like that.

He and his West Point friends believed in something called "strategy." That word was all over Washington D.C. in the first year of the war, very much as "four-dimensional chess" is now and is used in the same way, to signify a ridiculously complex, multi-step solution to a straightforward problem.

This is a common trap. Remember, cadets at West Point only got a few hours' training in military history and strategy, and what they drew from their West Point experience was an awed overestimation of "maneuver" warfare rather than striking the enemy directly, or bleeding him by occupying a strong defensive position. The very word "strategy", repeated by every fool in D.C. in 1861, drove Count Adam Gurowski absolutely crazy.

Gurowski had seen real wars, and knew that "strategy" meant speed, above all, speed and savagery. It did not mean what these fools thought it did:

> "They began to speak of strategy; plan to approach Baltimore on three different roads, and with about 35,000 men. Butler did it in one morning with two regiments!...[General Winfield] Scott... opposed the seizure of Alexandria. Is all [sic] that he knows of that hateful watchword—strategy—nausea repeated by every ignoramus and imbecile!" (*Diary. May, 1861*)

Gurowski had been in real wars. He'd lost everything and fled Poland at the point of Tsarist bayonets. And he knew that this supposedly "Napoleonic" buzzword, "strategy," meant speed, overwhelming force, and cruelty.

It made him furious (that is, more furious than usual) to hear D.C. bumpkins chanting the word "strategy" to mean avoiding battle, dreaming up counter-intuitive plans to "outflank" the enemy instead of meeting and destroying him. And it was especially galling to him to be lectured about it by so-called professionals, which in the US context meant West Point grads like McClellan.

Gurowski, like Forrest, believed in hitting first with overwhelming force. (An aside: I knew a woman who'd been a successful bully in elementary school. She'd never heard of Forrest, but she distilled his wisdom in an even more succinct formula. "Hit first" was her simple creed, and she said it worked like a charm.)

But alas, Gurowski was not in charge of the Army of the Potomac. Even alas-er, as it were, McClellan was. So the Peninsula Campaign was doomed from the beginning.

It should be noted that though the US Army performed very un-evenly in the first years of the war, the US Navy was uniformly efficient from the start.

So the first stage of McClellan's plan, the one involving ferrying his massive army to Virginia and disembarking near Richmond, went very well. The Army of Northern Virginia, commanded at this time by the Confederate Joseph Johnston, another West Point grad who specialized in retreats and withdrawals, was caught out of position to the north of Richmond, as McClellan's massive force popped up on shore far south and east.

A good commander would have force-marched the Army of the Potomac straight to Richmond before Johnston could come up. This would probably have ended the war and saved hundreds of thousands of lives.

McClellan, of course, was too much of a "strategy" man to do that.

The lead-up to the Seven Days was a game of feints by McClellan and Joe Johnston. But at the end of May 1862 Johnston was wounded and replaced by Robert E. Lee. Lee knew Northern Virginia better than a 21st-century real estate agent selling McMansions to DoD lobbyists. And more important, he knew McClellan.

Word had gotten around the small world of the Planter/West Point elite that Little Mac scared easy. So Lee dealt with McClellan in a simple, effective way—he attacked. That was all it took. Attack McClellan and he'd fold.

When Lee took over, the Union forces were a few miles from Richmond. A week later, they were in full flight, making for Harrison's Landing on the James River, where they could board ship to skedaddle by water. Despite all the talk about Lee's genius, there was nothing very complex about what happened in that disastrous week. It was simple: Lee attacked, McClellan retreated. Bury the dead and repeat.

McClellan never even saw those battles. He commanded from the rear, trying to manage the battles by telegraph. After all, it was the telegraph that made him The Young Napoleon so he stuck with it. And besides, he was, I think, a coward; plain, simple, outright physical coward.

This was not so common in his demographic, but if you look back at his first combat experience in Mexico, there's evidence for it. McClellan was assigned to ride with the "sappers," escorting cannon to the front. But he "was expecting trouble," and had equipped himself for it: "He carried a double-barreled shotgun, two revolvers, a saber, a rapier, and a bowie knife."[1]

Now that he was in command, McClellan stayed behind the lines, unlike most commanders. Johnston commanded from the front; that

---

1. Waugh, John C. *The Class of 1846: From West Point to Appomatox— Stonewall Jackson, George McClellan and Their Brothers*, New York, U.S.A., Ballantine Books, 1999

was how he'd gotten wounded. A. S. Johnson, the Confederate commander at Shiloh, was killed early in the battle. Lee pushed to be near the front line at least once, at the Battle of the Wilderness, when nervous soldiers, begged him to fall back—"General Lee to the rear!"[2]

But not McClellan. His command post was a few kilometers to the rear, usually in a comfortable house vacated by a civilian family, connected to the front by telegraph. He got only terse, partial summaries, and was often shocked when he finally rode up to the front (once the battle was over) to find that the situation wasn't so bad.

For a week, the pattern held: Lee attacks, McClellan falls back eastward, toward the James River.

Lee's advance wasn't cunning or brilliantly planned. He just attacked, attacked, attacked, and paid a nasty price. His Army of Northern Virginia took massive casualties, 20,000 dead, wounded, and missing in one week.

Lee's genius, if you can call it that, was simple: He knew McClellan, and he knew Little Mac would fold under pressure. Lee told his commanders, during the Seven Days Battles, "McClellan will make this a campaign of posts. He will take position [after] position, under cover of his heavy guns. I am preparing a line that I can hold with part of our forces in front, while [with] the rest I will endeavor to make a diversion to bring McClellan out."

That was his whole plan.

The heavy casualties Lee suffered have led some McClellan apologists to argue that he was playing four-dimensional chess via planned

2. "Lee to the Rear!" Robert K. Krick. 3/12/2025.
https://www.historynet.com/lee-to-the-rear/

withdrawals, counting on attrition a la Kutuzov. This is another "strategy" fallacy.

McClellan was just plain fleeing, doing what his adrenal glands were telling him. His subordinates were astonished to get the telegraphed command to retreat after every battle. They'd held their lines, moved close to Richmond, and were inflicting terrible casualties on Lee's forces. In short, conducting a successful campaign that Lee would have admired: strategically offensive, tactically defensive. When they got the telegraphed order to retreat, they were astonished. Why not remain in place and let the Confederates attack again and be bled out? But McClellan didn't have the guts. He was holding his breath in terror until he got his armies back in the shelter of the naval guns.

McClellan, with his 112,000-strong "bodyguard," reached Harrison's Landing early in July 1862. His soldiers fortified the place, and waited for their commander to tell them to retreat some more. Not that they needed to; the position was impregnable. Still trembling in this impregnable position, McClellan wrote his famous "Harrison's Landing Letter."

Here it is:

> "Mr. President: You have been fully informed that [the] rebel army is in [our] front, with the purpose of overwhelming us by attacking our positions or reducing us by blocking our river communications. I cannot but regard our condition as critical, and I earnestly desire, in view of possible contingencies, to lay before your excellency, for your private consideration, my general views concerning the existing state of the rebellion, although they do not strictly relate to the situation of this army or strictly come within

the scope of my official duties. These views amount to convictions, and are deeply impressed upon my mind and heart. Our cause must never be abandoned; it is the cause of free institutions and self-government. The Constitution and the Union must be preserved, whatever may be the cost in time, treasure, and blood. If secession is successful other dissolutions are clearly to be seen in the future. Let neither military disaster, political faction, nor foreign war shake your settled purpose to enforce the equal operation of the laws of the United States upon the people of every state.

"The time has come when the government must determine upon a civil and military policy covering the whole ground of our national trouble.... The responsibility of determining, declaring, and supporting such civil and military policy, and of directing the whole course of national affairs in regard to the rebellion, must now be assumed and exercised by you, or our cause will be lost. The Constitution gives you power sufficient even for the present terrible exigency.

"This rebellion has assumed the character of war; as such it should be regarded, and it should be conducted upon the highest principles known to Christian civilization. It should not be a war looking to the subjugation of the people of any State in any event.

It should not be at all a war upon population, but against armed forces and political organization. Neither confiscation of property, political executions of persons, territorial organization of States, or forcible abolition of slavery should be contemplated for a moment. In prosecuting the war all private property and unarmed persons should be strictly protected, subject only to the necessity of military operations. All private property taken for military use should be paid or receipted for; pillage and waste should be treated as high crimes; all unnecessary trespass sternly prohibited, and offensive demeanor by the military towards citizens promptly rebuked. Military arrests should not be tolerated, except in places where active hostilities exist, and oaths not required by enactments constitutionally made should be neither demanded nor received. Military government should be confined to the preservation of public order and the protection of political rights. Military power should not be allowed to interfere with the relations of servitude, either by supporting or impairing the authority of the master, except for repressing disorder, as in other cases. Slave contraband under the act of Congress, seeking military protection, should receive it. The right of the Government to appropriate permanently to its own service claims to slave labor should be asserted, and the right of the owner to compensation therefore should be recognized.

"This principle might be extended, upon grounds of military necessity and security, to all the slaves within a particular State, thus working manumission in such State; and in Missouri, perhaps in Western Virginia also, and possibly even in Maryland, the expediency of such a measure is only a question of time.... A system of policy thus constitutional and conservative, and pervaded by the influences of Christianity and freedom, would receive the support of almost all truly loyal men, would deeply impress the rebel masses and all foreign nations, and it might be humbly hoped that it would commend itself to the favor of the Almighty.

"Unless the principles governing the future conduct of our struggle shall be made known and approved, the effort to obtain requisite forces will be almost hopeless. A declaration of radical views, especially upon slavery, will rapidly disintegrate our present armies. The policy of the government must be supported by concentration of military power. The national forces should not be dispersed in expeditions, posts of occupation, and numerous armies, but should be mainly collected into masses and brought to bear upon the armies of the Confederate States. Those armies thoroughly defeated, the political structure which they support would soon cease to exist.

"In carrying out any system of policy which you may form you will require a commander-in-chief of the army, one who possesses your confidence, understands your views and who is competent to execute your orders by directing the military forces of the nation to the accomplishment of the objects by you proposed. I do not ask that place for myself. I am willing to serve you in such position as you may assign me, and I will do as faithfully as ever subordinate served superior.... I may be on the brink of eternity; and as I hope forgiveness from my Master, I have written this letter with sincerity towards you and from love of my country.

"Very respectfully, your obedient servant, Geo. B. McClellan, Maj.-Gen. Commanding. ("Headquarters, Army of the Potomac, Camp near Harrison's Landing, Va., "July 7, 1862.)

The strangest feature of this letter is that it only mentions military matters in passing, and only to paint a gloomy picture of imminent annihilation. It's as if McClellan isn't even interested in giving Lincoln an outline of the strategic situation. His interest is in dictating political terms to Lincoln.

The Young Napoleon had started composing his letter back when Joe Johnston, his fellow "strategy" devotee from the Point, was still commanding the Confederates defending Richmond. If Johnston

had remained in command, McClellan would have been at least be-
sieging Richmond, and would have remained a hero to the Northern
public. In that situation, he could have dictated terms to Lincoln. That
he could have done more than that definitely occurred to him more
than once. The idea of becoming President McClellan, Field Marshall
McClellan, King George the First, was one that he loved to bring up
in his letters home.

In one, written to his wife just three weeks after his Harrison's
Landing letter, he tells her:

> "I find myself in a new and strange position
> here—Presdt, Cabinet, Genl Scott & all deferring
> to me—by some strange operation of magic I seem
> to have become the power of the land. ... I al-
> most think that were I to win some small success
> now I could become Dictator or anything else that
> might please me—but nothing of that kind would
> please me—therefore I won't be Dictator. Admirable
> self-denial!" — George B. McClellan, letter to Ellen,
> July 26, 1861

He says he won't do it, but in a very un-reassuring way, the way a
cannibal in a snowbound cabin could say "I could just, y'know, whack
you and cook you... but that would be wrong. Oh come on, what're
you shaking for? Can't take a joke?"

And the scariest line of all is that smug self-congratulation: "Ad-
mirable self-denial!"

By late July 1862, McClellan was sure Lee would not chase and destroy him, so he was free to indulge the boastful side of his simple bipolar world.

When he wrote the Harrison's Landing letter on July 7, he was skulking at the other pole, sheer terror. McClellan had composed this extraordinarily strange letter at two different times. The part containing his sermon to Lincoln about how the war should be conducted was first written when the advance on Richmond was going well. It's clearly the product of McClellan's mind enjoying those dreams of kingship that he coyly mentions in the letter to his wife. All that changed when Lee took command and began attacking. The letter was finished while McClellan was cowering in terror after retreating to Harrison's Landing, dreaming of some self-indulgent martyrdom at Lee's hands. The result is a pompous, craven screech, one of the most repulsive texts you'll ever read. No one in their right mind, looking over this psychotic screed, would have sent it. So of course Little Mac sent it.

He tells Lincoln the bad news immediately: "[the] rebel army is in [our] front, with the purpose of overwhelming us by attacking our positions or reducing us by blocking our river communications. I cannot but regard our condition as critical..."

You would think this would be the main topic. But no, it's just a chatty introduction, breezily informing the President that his great army is about to be destroyed.

Of course it's nonsense in military terms; the Army of the Potomac outnumbered Lee's Army of Northern Virginia throughout the Peninsula Campaign, and it's hard to see how the Confederacy could have "block[ed Union] river communications" when the US Navy controlled the estuaries completely. There were a half dozen

USN gunboats within McClellan's view, on the James River, if he'd bothered to look.

The point of this melodrama is to give the remainder of the letter the force of a deathbed declaration, a last wish by the soon-to-be martyred leader. McClellan seems to know this, admitting in the first paragraph that his political sermons "do not strictly relate to the situation of this army or strictly come within the scope of my official duties."

This transgression is supposed to be excused because he's about to be killed, or so he claims. Remarkably adolescent. It's hard to imagine grownups like Grant or Thomas writing, "I'm about to be slain, SLAIN I tell you, and so with my last breath I beg you not to mess with slavery."

Because that's what the rest of the letter is about: preserving slavery.

It's important to notice how little McClellan cares about military matters, and how intensely he cares about preserving "property relations," meaning slavery. A lot of American historians have been reluctant, let us say, to notice this, but it's the most glaring feature of the entire letter.

The transition to this non-military topic is a strange one, starting toward the end of the first paragraph and continuing for another three melodramatic paragraphs:

> "Let [nothing] shake your settled purpose to enforce the equal operation of the laws of the United States upon the people of every state. The time has come when the government must determine upon a civil and military policy covering the whole ground of our national trouble. The responsibility of determining, declaring, and supporting such civil and military pol-

icy, and of directing the whole course of national affairs in regard to the rebellion, must now be assumed
and exercised by you, or our cause will be lost. The
Constitution gives you power sufficient even for the
present terrible exigency. This rebellion has assumed
the character of war; as such it should be regarded, and
it should be conducted upon the highest principles
known to Christian civilization."

McC. is taking his time here, dancing around the point. His first
allusion to it is very oblique: "... the equal operation of the laws...
upon... every state."

One of the biggest cliches of the Southern grievance was that their
beloved Fugitive Slave Act was being thwarted by Abolition states
like Massachusetts—after they'd gone to so much trouble to ram it
through Congress. That's what the reference to "every state" means:
Force New England states to stop interfering with the "recovery" of
escaped slaves.

Then comes the really weird part. First, McClellan tells the president that it's his, the president's, responsibility to "determine upon a
civil and military policy," as if that's a new development. It was always
the president's responsibility.

What McClellan is doing here is kicking his military problem, his
ridiculous conviction that he was about to be overrun, up to Lincoln,
as if Lincoln's policies are what will determine the outcome of the
upcoming battle. If McClellan had won outside Richmond, he could
have forced Lincoln's hand on this, "negotiating from a position of
strength," to use one of Kissinger's favorite cliches. But as it turned
out, McClellan is now trying to force Lincoln's hand from a position
of weakness.

So there's an oddly confused sound to his counsel. First, "this rebellion has assumed the character of war... "—which I simply don't get. Didn't it "assume [that] character" when South Carolina troops fired on Fort Sumter? Perhaps this distinction between "rebellion" and "war" is a 19th century legalism of the sort favored by right-wingers, some analogue to the current cliché, "We're NOT a democracy, we're a REPUBLIC."

Whatever it means, "war" certainly seems like the more serious term. So it's odd that the sentence continues, "... it should be conducted upon the highest principles known to Christian civilization."

Perhaps McClellan meant that "rebellion" can be suppressed ruthlessly, whereas "war" must proceed on more gentlemanly lines. The European tradition would certainly suggest this (creaking of gallows, soft sighing wind-noises).

It's so badly written you can't say for sure. All you can be sure of is that the point of the letter is to save the Planters, save slavery, and lobby for a soft war. The reason it doesn't work (and I feel like I'm giving McClellan a posthumous lesson in first-year Comp here) is that he didn't revise his draft properly. You can't write a letter demanding policy changes when you're winning, and then add an intro saying "By the way we're about to be overwhelmed and I'm going to be killed" and then resume your arrogant list of demands to the President. Just plain sloppy.

How would it strike any sane reader to hear that (a) McClellan's army is about to be overrun, and he himself is "on the brink of eternity"; (b) the rebellion is therefore officially a war; and therefore (c) this war "should be conducted upon the highest principles known to Christian civilization."

That just doesn't track. (a) and (b) describe an apocalyptic scenario, so it's a strange non sequitur the conclusion, (c) is "therefore, let's play nice."

This paradox gets unravelled very quickly as McClellan explains what those "highest principles" really are. McClellan lists these "principles" in ascending order, in a gradatio starting with the severest ("military executions") and ending, ludicrously, with the prohibition of "offensive demeanor" toward Confederates.

If Private Frost of Aliens were to ask McClellan "What're we supposed to use, harsh language?" the answer would be, "Certainly not, how dare you!"

Finally, Little Mac gets to the point, focusing on slavery at great length:

> "Military power should not be allowed to interfere with the relations of servitude, either by supporting or impairing the authority of the master, except for repressing disorder, as in other cases. Slave contraband under the act of Congress, seeking military protection, should receive it. The right of the Government to appropriate permanently to its own service claims to slave labor should be asserted, and the right of the owner to compensation therefore should be recognized."

The First Confiscation Act had already passed Congress. It became law on August 6, 1861. It allowed Union commanders to "confiscate" escaped slaves. This legal twiddle was invented by Benjamin Butler, who reasoned that if the Planters insisted that their slaves were mere property, then that property could be confiscated like hay or bacon or cotton bales by Union troops if this "property" was being used to help the rebellion. And, as every combat officer in the Union Army knew, slaves were the Confederacy's biggest asset. They did the work, they brought in the crops, they built the fortifications that the CSA fought behind. It should have been obvious to all that they should be taken from their "masters."

So McClellan can't argue against the right to confiscate slaves. Instead, he tries to persuade Lincoln to offer compensation, that is to pay the "owners" for the slaves, despite the fact that the owners were in open revolt against the USA.

West Pointers like Buell and McClellan ordered their troops not to let escaping slaves within Union lines. The First Confiscation Act forced them to change that policy, but they could always argue that the slaves they rejected were not part of the Confederate war effort.

So the radical Republicans pushed through a second Confiscation Act which allowed confiscation of all rebel property, no matter what its supposed use was. That second act was about to be made law when McClellan wrote. He's trying to find a way to squirm his "don't touch slavery" stance into alignment with the two confiscation acts.

He claims this "Christian" policy "would deeply impress the rebel masses and all foreign nations." But would it have? Officers in the field were unanimous: Confederate civilians could not be wooed. Their hearts and minds were already spoken for. They were raising crops, or rather their slaves were, to feed their male relations in the CSA.

As for "foreign nations," the truth was just the opposite: it was slavery that prevented Britain and France from intervening on behalf of the South. The British elite would have loved to back a weak, amateurish oligarchy like the Planters. They would have driven that wedge in deeper, decade by decade, until the USA was broken forever as a potential rival.

US diplomat Cassius Clay wrote of the British elite:

> "They hoped for our ruin! They are jealous of our
> power. They care neither for the South nor the North.
> They hate both."

But there was a middle class in Britain by 1860, and whatever their blind spots (don't get me started), they would not accept outright slavery in an ally.

McClellan couldn't see it because he surrounded himself with reactionary, aristocratic foreigners who had a "conservative" view. With his massive, wounded ego, McClellan was the perfect snob, and his preference in foreign friends showed it.

Try to imagine McClellan listening to someone like Gurowski. It's impossible, of course—McClellan would have looked at Gurowski's sloppy, obese, badly-dressed appearance and dismissed him even before Gurowski had a chance to start spouting his "radical" ideas. But if any of you are thinking of writing absurdist comedy, it's a great idea for a story. DC was a small town, so Gurowski and McClellan probably passed each other in the street a few times, Gurowski toiling along on foot, sweating and cursing the idiots around him, McClellan traveling in state, striking one of his endless poses...

McClellan preferred his foreigners more snobby and "conservative," and there's nobody more like that than the aristocratic detritus of 1848. So he surrounded himself with surplus Orleanists, Prince de Joinville and his nephews. He could show off his fluent French with them, though de Joinville himself was "profoundly deaf." Which, come to think of it, opens more glorious comedic vistas.

Anyway, McClellan was just as deluded about Europe as he was about the US, which is saying something. He knew nothing about the hard-won, powerful revolutionary movements that had overthrown his French aides' father a few years before the US Civil War. Of course these "foreign nations" don't really matter much. While the Confederacy needed foreign support, the Union didn't need or even want Europe involved. The Union navy controlled the seas around its own borders as well as the Confederacy's and had all the population, manufacturing, and raw materials it needed. McClellan only mentions foreigners' opinions in defense of slavery, his real topic.

He gets to the point at last, saying that if Lincoln emancipates the enslaved, the Union armies will dissolve:

> "A declaration of radical views, especially upon slav-
> ery, will rapidly disintegrate our present armies."

Everyone in Little Mac's clique repeated this talking point. In every letter he wrote, McClellan's Wormtongue, Fitz John Porter, was re-peating the lie that Emancipation would fracture the Union forces. Here's one from July 1862, the same time that McClellan was cower-ing at Harrison's Landing:

> "The [Emancipation] proclamation was ridiculed in
> the army—caused disgust, discontent, and expres-
> sions of disloyalty... amounting, I am told, to insub-
> ordination. [It will] prolong the war by rousing the

bitter feelings of the South... while the reverse with us." —F. J. Porter

Porter added gleefully that the Union forces were "doomed," recounting the horrible fate in store for each Union command, all due to this terrible Proclamation.

It's as if he was hoping for a Union defeat.

Porter, Buell, and McClellan may even have believed their line that the Confiscation Acts and the Emancipation Proclamation would rupture the Union forces, but that just goes to show that they needed to get out more. Soldiers might have been ambivalent about emancipation, but they were solidly in favor of "confiscating" escaping slaves. A few Border-state officers quit in a rage; a lot of enlisted men grumbled for a few days; and that was all.

On the other hand, many mid-level officers were gratified by the Confiscation Act. They'd been furious that they weren't allowed to shelter escaping slaves. A few of them felt this way because they were genuine abolitionists, but most just saw the obvious fact that slavery freed up the very Confederate soldiers who were shooting at them. Deprive the Confederacy of slave labor and it would have no food, no fortifications, no hope at all.

Line officers like Col. John Beatty knew this:

> "The policy we need is one that will march boldly, defiantly, through the rebel States, indifferent as to whether this traitor's cotton is safe, or that traitor's negroes run away... Slavery is the enemy's weak point, the key to his position. If we can tear down this institution, the rebels will lose all interest in the Confed-

eracy, and be too glad to escape with their lives, to be very particular about what they call their rights. " *(The Citizen-Soldier "July, 1862")*

But McClellan had no contact with Republican mid-level officers like Beatty. Sticking to his proto-Confederate West Point clique, he assumes everyone in the Union Army felt like they did. After presenting his bizarrely arrogant demands for policy change, McClellan switches to a grovelling tone. He concludes by offering "... to serve you [Lincoln] in such position as you may assign me, and I will do as faithfully as ever subordinate served superior."

This is a good ol' flat-out lie. Little Mac was writing letters to his wife, the long-suffering Ellen Mary, in which he called Lincoln "the original gorilla." Lies come very easily to the "conservative" generals, which really should be taken more seriously by CW historians who trust these generals when they proclaim their loyalty.

And after all this conniving bombast, McClellan reaches down for his big aria, the very operatic closing of the letter:

> "I may be on the brink of eternity; and as I hope forgiveness from my Master, I have written this letter with sincerity towards you and from love of my country."

No other CW general could have written this purple paragraph. "On the brink of eternity"?! McClellan is actually imagining himself shot or bayoneted by an overwhelming Confederate force.

Look at a map of Harrison's Landing shows you how insane he had to be to write this. Not to mention that if he was going to be killed, his entire Army would have to be killed first, since he always chose the most protected spots for his command posts.

But the soldiers' fate doesn't seem to have struck him. It's all his glorious death scene, imagined and avoided.

He can't stop lying, even while emoting on his imaginary deathbed. "Sincerity" is precisely what this letter is not written with. It's something much creepier than standard insincerity, though—something more like hysterical self-pity with the remnants of rodent cunning.

You see the cunning in his capitalized "Master," meaning God. That word occurs earlier in the letter, when he urges Lincoln not to interfere with "the authority of the master"—without a capital "M" this time—implying the naturalness, the permanence of the slave/master relationship.

McClellan was not in control of his rhetoric—or, in fact, his mind—at Harrison's Landing. He was considering becoming a dictator, as he himself wrote. Until recently, U.S.C.W. historians have had a sentimental attachment to the idea of America as a place that was immune to military seizures of power. We don't have military takeovers. We don't have outright traitors in the top echelons of the armed forces. So, in their estimation, McClellan must have been just plain incompetent, because he couldn't have been a traitor.

When they tried to contextualize the sinister conduct of generals like McClellan, Buell and Porter, they ruled out the possibility of disloyalty right at the start. In the accounts of mainstream Civil War historians, you will always find a disclaimer to the effect that "Of course General Buell [or McClellan or Porter] was a true patriot, devoted to the Union cause, despite his many mistakes."[3]

Those growing up in the 21st century will have a hard time believing Americans were so naïve. But when it came to judging American generals, they really were. Most-20th century histories of the Civil War assume an American exceptionalism to the point of explaining apparent treachery as just about anything else: incompetence, mental illness, pettifoggery. Drop the exceptionalism, and you have a great deal of evidence that McClellan for one was very close to becoming Generalissimo and was only prevented from doing so by military failure and his own cowardice. You have to wonder, what would have happened if McClellan had managed to take or even just besiege Richmond?

---

3. For an example, see an otherwise good book, Stephen Engle's *The Most Promising of All*, about Buell.

Chapter Eight

Worst Battle Speech Ever

The armies that were assembled in the first year of the war worked with many disadvantages: an arrogant yet inexperienced officer corps, soldiers who were not willing to be treated like livestock in the European manner. Under these circumstances, every regiment worked through its own family drama in the first year of the war, coming to terms with the need for subordination, the development of an etiquette that would be acceptable to volunteer soldiers, and the even more pressing need to rid itself of dangerously incompetent officers. Here's an example of how that happened from one of the finest memoirs of that time.

We have so many great voices in the Radio War Nerd community that every now and then, a subscriber's comment calls for more than a short social-media response. That's how I felt reading Ben Hosen's Facebook post on Bill Paxton's famous "Game over! We're fucked!" speech in *Aliens*:

"Nothing but love for Bill Paxton (RIP) but in *Aliens* wasn't he the worst infantry officer ever? "We're all gonna die, we're fucked" isn't the most inspiring thing to hear over and over again. Also very problematic towards his Latina squadmate."

Of course, RWN guest Gabriel Uriarte raised the proper objection:

"'Officer'!? What kind of war nerd are you? He was a corporal commanding a squad (the other one was Hicks), an NCO of not a very high rank. And I don't think Hudson is even trying to make sergeant ("I was getting short" not a comment someone who planned to make a life in the Corps would make, and cracking wise at platoon briefings not advisable either). Only officer in that outfit was, infamously, Gorman (pilot may have been a warrant officer, but that's a different animal)."

To which Ben replied:

"As a loyal Russian bot I do not understand the concept of NCOs."

*

Now, that comment sequence is quality War Nerd banter.

And yet, while savoring it, I couldn't help wanting to make a pedantic point that there HAVE been pre-battle speeches IRL, or rather IRW, which were even more depressing than Paxton's.

Since the original posts scored high on pedantry themselves, you might call my much longer response here meta-pedantry, and who among us could resist the chance to meta-pedant? I never meta pedant I didn't try to top, hur hur.

The best of all defeatist speeches comes from Colonel John Beatty's USCW memoir, *The Citizen-Soldier*. I've mentioned this book many times, as the best USCW memoir ever written, but I've never had an excuse to introduce Beatty's depiction of Colonel Isaac Morrow, who commanded Beatty's volunteer regiment (3rd Ohio) in the first year of the war.

There was a good bit of confusion in that first year, and part of it came from officers trying to figure out how to talk to volunteers, or "citizen-soldiers." The hardcore West-Pointers, like Don Carlos Buell, went with the traditional view, that you simply did not talk to enlisted men at all. In the European armies on which the USCW forces were modeled, soldiers were considered beasts, to be restrained or coerced as needed, but never to be addressed as fellow humans.

Other West Point grads, like McClellan, courted the enlisted men, but at a distance, issuing occasional gung-ho speeches about their glorious prospects. Still other regular officers, like George Thomas and Grant, took what was probably the best approach, not cozying up to the ranks but doing their best for their troops.

Then there was the officer Beatty served under in his first campaign. Beatty usually calls him "The Colonel" but occasionally slips and mentions his name Isaac Morrow, one of the war's many command failures.

Beatty, a wonderful narrator, dislikes The Colonel from the start, hinting, in an entry for July 1861, that Morrow is a braggart, a drunkard, and a fool. Among his other failings, Morrow gets hysterical at the prospect of a battle:

> "...Colonel Marrow [sic], apparently under a high
> state of excitement, informed me that he had just had
> an interview with George (he usually speaks of Mc-
> Clellan in this way), that an attack [on the Confeder-
> ate forces] was to be made, and that the Third [Ohio]
> was to lead the column."

I can't swear that Beatty's phrase, "high state of excitement" is meant to imply that Morrow was drunk when he talked to Beatty, but it seems likely given Beatty's many references to Morrow's long absences in town. "High state of excitement" seems to be Beatty's euphemism for "dead drunk."

The Colonel comes back from McClellan's HQ, routs Beatty out of his tent at midnight, and orders him to assemble the men of the regiment. Beatty sets the scene, as these new volunteers, with no combat experience, stumble out of their blankets:

> "They looked like shadows as they gathered in the
> darkness about their chieftain. It was the hour when
> graveyards are supposed to yawn, and the sheeted dead
> to walk abroad."

Beatty was expecting The Colonel to give the men some form of encouragement as they prepare for their first fight. Such speeches, a venerable genre in the ethopoeia tradition, were expected of commanders before a battle. Tacitus's most famous phrase, "They make a desert [i.e. a wasteland] and call it peace," comes from a speech Tacitus places in the mouth of a German chieftain readying his followers for battle against the Romans.

Most of these speeches are standard pep talks. A few are remembered as especially effective, like Julius Caesar's address to the troops before the Battle of Pharsalus (48 BC).

Schoolboys were trained to read, memorize, and compose such speeches for more than two thousand years. Boys who went to school in London, Paris, or Rome would have learned such speeches in the original Greek or Latin. But Beatty, Ohio-born and -raised, probably only saw English paraphrases in his schoolbooks. But, one way or another, he would have had some familiarity with the tradition of the glorious pre-battle oration.

So when the Colonel assembles the regiment at midnight, Beatty expects something on these traditional lines, inspiring the men with pride in their own prowess, hatred of the enemy, or a sense of their responsibility to the cause.

He soon learns he was wrong:

> "The gallant Colonel, with a voice in perfect accord with the solemnity of the hour and the funereal character of the scene, addressed us..."

Beatty then gives the Colonel's speech, or at least his reconstruction of it. This is one of the funniest scenes in *The Citizen-Soldier*. You tell me, does this speech put Paxton's speech in *Aliens* ("Game over, man!") in its place or not?

The Colonel begins:

"Soldiers of the Third: The assault on the enemy's
works will be made in the early morning [of July 11,
1861]...The Third will lead the column. The secessionists [i.e. Confederates] have ten thousand men
and forty rifled cannon. They are strongly fortified.
They have more men and more cannon than we have.
They will cut us to pieces."

That last sentence must have been the moment that Beatty and the
other subordinate officers realized something was very wrong. You, the
commanding officer, might make an inspiring speech to your troops
that mentions the enemy's superior numbers, maybe by way of saying
something on the lines of *"Somos pocos pero locos"*—but you do not,
under any circumstances, say "They will cut us to pieces." That's
definitely not in the ethopoeia manual.

And it only got worse from there. Maybe The Colonel had entered
the weepy phase of drunkenness, or maybe he was just plain scared,
but he keeps on making things worse with every phrase, saying:

"Marching to attack such an enemy, so intrenched
[sic] and so armed, is marching to a butcher-shop
rather than to a battlefield. There is bloody work
ahead. Many of you, boys, will go out who will never
come back again."

Beatty describes his reaction:

"As this speech progressed my hair began to stiffen at
the roots, and a chilly sensation like that which might

ensue from the unexpected and clammy touch of the
dead, ran through me."

With a classic litotes, Beatty says that, after being told they'll be
slaughtered next morning, that "[t]heological questions which before had attracted little or no attention, now came uppermost to our
minds." The enlisted men try to improve the mood by stirring up the
campfire, but "...the little blaze which sprang up revealed scores of
pallid faces. In short, we all wanted to go home."

Beatty reviews his memory of other great commanders' pre-battle
orations at this point:

> "When a boy I had read Plutarch, and knew some-
> thing of the great warriors of the old time; but I could
> not, for the life of me, recall an instance when they
> had made such an address to their soldiers on the eve
> of battle. It was their habit, at such a time, to speak
> encouragingly and hopefully."

So, bolstered by his memory of classical ethopoeic speeches, Beatty
takes the risky step of trying to stop the Colonel:

> "...I plucked him by the sleeve, took him [to] one
> side, and modestly suggested that his speech had had a
> rather depressing effect on the regiment...I urged him
> to correct the mistake, and speak to them hopefully."

You'd think that even the hammiest tragedian would take a little direction at this point, but The Colonel wasn't having any of Beatty's constructive criticism:

> "He replied that what he had said was true, and [the men] should know the truth."

The punchline here is that it wasn't true, none of it. When the grim day dawns,

> "...instead of being called upon to lead the column, we were left to the inglorious duty of guarding the camp, while other regiments moved forward toward the enemy's line."

What ended up happening that day, The Battle of Rich Mountain, was a Union victory, an almost bloodless one. But it was, as military historians love to say, a costly one. Not in terms of casualties, but because it made McClellan a star. He was in overall command, but his subordinate, Rosecrans, a far better general, won the battle by finding a path through the mountains to show up in the Confederates' rear.[1]

---

1. I know what happened to Rosecrans after Chickamauga, but I also remember what he did in the Tullahoma Campaign, one of the few cases where the "War of Maneuver" so worshipped by West Point grads actually worked. He deserves more credit than he gets for that one.

At that point, McClellan, with the main force in front, was supposed to attack and catch the Confederates in the classic pincer movement. But McClellan, being McClellan, lost his nerve.

McClellan's career survived his first flinch, but the Colonel's midnight oration finished him as a leader of men. Beatty says, in a diary entry soon after the battle, that:

> "The men repeat his Rich Mountain speech with slight variations: 'Men, there are ten thousand secessionists in Rich Mountain, with forty rifled cannon, well fortified. There's bloody work ahead. You are going to a butcher-shop rather than a battle. Ten thousand men and forty rifled cannon! Hostler, you d----d scoundrel, why don't you wipe Jerome's nose?' Jerome is The Colonel's horse."

After his humiliation, the Colonel becomes even more drunk and derelict:

> "The Colonel spends most of his time in Louisville. He complains bitterly because company officers do not remain in camp, and yet fails to set them an example..."

And a few days later,

> "The Colonel has practically abandoned the regiment in this period of bad weather..."

And my favorite Beatty-ism:

> "We hear of the Colonel occasionally. He is still at
> Louisville, running his train on the broad gauge."

By January 1862, the Union forces were tightening up their standards for regimental commanders. Beatty puts all his complaints about Colonel Morrow in a letter, allows Morrow to read it before sending it, and offers him a chance to resign, and presumably ride home on Jerome's back without the disgrace of being officially removed:

> "[The Colonel]...tore open the envelope, and seated
> himself while he read. In less than an hour [two officers] called on me to report that The Colonel would
> resign if I would withdraw the charges. I consented to
> do so."

You can sense Beatty's anxiety as the Colonel delays his exit:

> "[The Colonel] passed through the...quarters, shaking hands with the boys and bidding them farewell.
> Still later he made a speech, in which he called God to
> witness that he was a loyal man, and promised to pray
> for us all. The regiment is disorderly, if not mutinous.
> The best thing [he] can do for it and himself is to get
> out."

You can infer from Beatty's anxiety that The Colonel's "valedictory speech" was much more effective than his pre-battle oration. Soldiers generally prefer fond goodbyes to predictions they're all going to die, and in the first year of the war, many units preferred weak commanders who flattered them to phlegmatic officers like Thomas or Beatty.

McClellan is the supreme example; most of his soldiers truly liked him, though they slowly came around to the realization he was no tactician.

The Colonel's story ends like a miniature version of McClellan's, though with a lot more alcohol involved:

> "January 8 [1862]: The Colonel has bidden us a final
> adieu. His most devoted adherents escorted him to the
> [railroad] depot, and returned miserably drunk."

If Beatty had been writing a screenplay, his resolute stand against the drunken Colonel Morrow would have earned him the regiment's respect. It didn't work out that way.

Beatty got one letter of support after the Colonel staggered onto his homeward train. Just one. The rest of the regiment hated his guts. He drilled them, made them stay in camp when the lights of town were twinkling, and generally made himself unpopular.

In the same month he forces the Colonel out (January 1862), Beatty gets in an argument with a drunken soldier who is trying to bring a couple of canteens full of whisky into camp. The soldier has no fear of being sent to the guard house, so Beatty comes up with a new punishment, ordering the man tied to a tree with a saddle girth. The regiment doesn't like this at all: "It was [in their view] a high-handed outrage upon the person of a volunteer soldier; the last and worst of the many arbitrary and severe acts of which I had been guilty."

This leads to open mutiny:

> "The regiment seemed to arise *en masse*, and led on by a few reckless men who had long disliked me, advanced with threats and fearful oaths toward my tent. The bitter hatred which the men entertained for me had now culminated."

Beatty ends up facing most of the regiment alone, pistol under his coat, daring them to untie the drunken soldier. They back down—they're not quite willing to kill an officer, but it's a close thing. Beatty, usually a cheerful diarist, says, "I feel greatly discouraged."

Well, it'd be enough to discourage almost anybody, seeing hundreds of people staring at you with hatred. Reminds me of my time teaching med. students at Otago University, but I digress.

What you see in Beatty's memoir is the chaos of the Union Army in the first year of the war, as everybody struggled to work out the relationships between officers and enlisted men. Many of the top brass hated the whole idea of volunteer soldiers, who were obstreperous and demanding.

General Don Carlos Buell despised his troops largely because most of them were volunteers rather than regulars, and they returned the favor. At the middle ranks, like that of the Colonel, the officers who'd been elected by "the boys" veered in the other direction, drinking with the ranks, refusing to enforce any rules that annoyed the men, and, as at Rich Mountain, favoring them with very frank previews of upcoming engagements, no matter how dire.

That wasn't a good move, and neither was Buell's cold loathing.

So, the relationships were being negotiated, often violently, during the first year of the war. This negotiation played out dramatically in Beatty's regiment. After the confrontation over the tied prisoner, the regiment obeys orders sullenly. Their attitude only changes at the Battle of Perryville (October 8, 1862) when Beatty had his men lie down under heavy fire while he paces back and forth, refusing calls to lie down himself.

Perryville was one of the bloodiest battles of the war, per capita: 8,000 casualties out of 36,000 troops engaged—a quarter of those involved were killed, wounded, or captured. The battle consisted of units blasting away at each other among the Kentucky hills, without much fancy maneuver.

It should have been an easy Union victory, because the Union had 55,000 men available, but Buell, the Union commander, sat in his tent reading a book, refusing to believe the reports of his despised volunteer troops that they were under attack. As a result, only 22,000 Union troops fought, while another 33,000 waited for orders. (Buell's excuse was that he was in an "acoustic shadow," unable to hear the noise of battle. I don't buy it from him or from Grant at Iuka either.)

Buell's Olympian scorn was discredited by that performance and he was booted from command, marking the end of the old-school attitude that mere enlisted men were beneath contempt. Beatty's sins as a rule-crazy martinet were instantly forgiven, because he was there on the battle line, and had been offering himself as a target all day.

Beatty says that under Buell's ultra-cautious maneuvers, "the boys" had been spoiling for a fight, but Perryville cured them:

> "They have long sought for a battle, and often been disappointed and sore because they failed to find one;

but now, for the first time, they realize what a battle
is."

Reevaluating hated officers was part of the epiphany. Beatty had
proved himself as far as the regiment was concerned. After the battle,
he gets the word from a fellow officer:

> "Meeting Captain Loomis yesterday, he said, 'Do you
> know you captured a regiment at Chaplin Hills [Per-
> ryville]?' / 'I do not.' / 'Yes, you captured the Third
> [Regiment.] You have not a man now who wouldn't
> die for you.'"

OK, allowing for Victorian sentimentality (especially where extreme gore and bathos can be mixed) and granting that the regiment
probably wasn't quite as starry-eyed as Beatty claims, it's plausible that
his display of heroic death-wish at Perryville did change his soldiers'
minds about him. The Army, especially in the West where Beatty
fought, had had its shakedown cruise in the first year of the war, and
was coming to a sort of consensus about what officers could do, and
soldiers would take, in making war effectively.

Chapter Nine

Arm The Slaves? The Confederacy Would Rather Die

A s *Sherman realized, after the strategic disasters of 1863, the Confederacy was no longer fighting to win. They were simply fighting to continue fighting. Nothing shows this more clearly than the strange fate of the attempt to shock Confederate leaders into using their one strategic resource: the enslaved.*

"Set Uriah in the forefront of the hottest battle, and retreat from him, that he may be struck down and die." *(2 Samuel 11)*

By 1864, the Civil War had turned against the Confederacy so clearly that even the Fire-Eaters, the hardcore secessionists, had to admit that The Cause was in deep trouble. Early in the war, the Union and the Confederacy each had about 200,000 soldiers, but the war was so much bigger and bloodier than most expected that both sides had to recruit fresh bodies and put them on the front line. The Union, which could draw on a population of 19 million, could replace its losses, though the process wasn't pretty.

The Confederacy, with a much smaller population of 5.5 million citizens, could not. There were about 9 million people in the Confederacy at the beginning of the war, but 3.5 million of them were slaves, not citizens. There were only 5.5 million classed as "white" and thus citizens eligible for military service.

For the Confederate elite, it went without saying that only whites could be soldiers. Or at least it did until things got rough. There were free black people in the Confederacy, but they were classified as a distinct group for conscription purposes, forced to enlist as laborers but excluded from combat. Even when free black people volunteered to fight for the Confederacy early in the war, they were rebuffed:

> "Despite the initial enthusiasm of New Orleans's free men of color for the Confederate cause, both state and national Confederate officials were uncomfortable with the idea of black soldiers within their ranks."

Enslaved black people were certainly used by Confederate armies: as laborers building the fortifications the Confederate armies fought behind; as "personal servants" to wealthy Confederate soldiers; as teamsters, driving supply wagons; as forced labor in the fields, keeping the farms going and feeding the wives and children of their supposed "masters." No Confederate had any problem with this sort of employment for enslaved people.

In fact, the only people who had a problem with it were the "owners" of enslaved people commandeered by Confederate forces. Time after time, the diaries and letters of Confederate officers fume at the difficulty of getting slaveowners to lend their slaves to the Confederate military. Over and over, soldiers say bitterly that plantation owners willingly sent off their own sons to fight and die for the Confederacy, but would not lend their enslaved laborers unless forced to do so.

> "The sanctity of slave property in this war," admitted Confederate Secretary of War James A. Seddon, "has operated most injuriously to the Confederacy..."

But the idea of giving the enslaved guns, making them into soldiers, was anathema. Howell Cobb, a Georgia politician, summed up the objections: "If slaves will make good soldiers our whole theory of slavery is wrong—but they won't make soldiers."

Cobb wrote that sentence in a letter to James Seddon, the Confederate Secretary of War. The sentence is the most concise possible summary of the Confederate dilemma. You can feel the vertigo, the disjunction, between the two clauses divided by the dash.

Cobb first peeks at the appalling possibility: "If slaves will make good soldiers..." and then draws back in horror: "...but they won't make soldiers."

This is a common form of argument, more common than people want to admit. You sketch the "appalling vista" which undercuts your whole shared worldview, counting on your audience's recoil to repel it, without the need to come up with a good reason. Cobb says plainly in an earlier paragraph of this letter that he rejects the idea of enlisting black soldiers because it's simply too "pernicious":

"I think that the proposition to make soldiers of our slaves is the most pernicious idea that has been suggested since the war began...My first hour of despondency will be the one in which that policy shall be adopted."

Other Confederates called the idea "monstrous," "revolting," and claimed it was "at war with...social, moral, and political principles." No one said it wouldn't work, or that there were viable alternatives. Racial hierarchy was just more fundamental to them than the survival of the Confederacy.

This big flinch kept the Confederacy from using its only possible source of new soldiers in the later stages of the war. And damn, did they need new soldiers. In fact, it's stunning that Cobb said, in January 1865—a few months from the inevitable collapse of the Confederacy—that he would only feel "despondency" if the Confederacy enlisted former slaves, implying that even the collapse of the Confederacy would be less traumatic than freeing and enlisting slaves.

Arming the enslaved had always been the Confederacy's biggest nightmare. Even at the beginning of 1863, when the Confederates were winning battles (at least in the East), the Emancipation Proclamation triggered the Confederate Senate so intensely that they fast-tracked a law ordering that:

"Every white person who shall act as a commissioned or non-commissioned officer, commanding negroes or mulattoes against the Confederate states, or who shall arm, organize, train, or prepare negroes or mulattoes for military service, or aid them in any military

enterprise against the Confederate States, shall, if captured, suffer death."

Now, y'might say, "Well sure, the Confederate elite would naturally get hysterical about former slaves fighting for the Union, but
they might be more reasonable about enlisting them to fight for the
Confederacy."

Well, you'd be wrong. As furious as they were about ex-slaves fighting for the Union, the notion of using former slaves as Confederate
soldiers upset the Confederate elite even more.

But by 1864, they were just plain running out of white men. By the
middle of the year, the Union had about 600,000 men in uniform. The
Confederacy had to work very hard to keep about 200,000 soldiers in
its armies, one-third the Union force.[1] Since they refused to consider
using non-whites as soldiers, the Confederate elite fell back on getting
every white man into uniform. Conscription was extended to all white
males between 15 and 50, then up to 65. Home Guard units hunted
the woods for deserters. Jefferson Davis harped endlessly about the
need to conscript all white men, as if there were vast numbers of them
hiding out in attics.

But there weren't. The grim truth was that, unless they were willing
to enlist ex-slaves, the Confederates just didn't have the manpower
(gendered term used advisedly; no one seems to have entertained the
notion of female soldiers at any point in the war).

---

1. See the graph "Strength" at the National Park Service "Civil War
Facts" website https://www.nps.gov/civilwar/facts.htm Accessed
March 12, 2025.

The Confederacy had resorted to conscription earlier than the Union, and no matter how Davis harped on squeezing slackers into the field, the truth was that the Confederacy was already over-mobilized. Deserters were hunted down and shot from 1862 on. One Confederate wrote that, as early as the battles of Corinth, "...some men had to be shot, merely for discipline's sake... Men were shot by scores..."

As the Confederates' winning streak ended in the summer of 1863, the problem of desertion and "apathy" got worse.

By 1864, many Confederate units had a back-line, consisting of up to 10 percent of all troops, whose job it was to follow troops into battle and shoot anyone who tried to run away. Sometimes units from western states like Texas and Arkansas, still gung ho, would warn units from wavering states like Georgia or North Carolina that they would shoot down anyone who tried to run:

> "Some of our boys...go to the breastworks in front of us to see what soldiers are there, because they have no confidence in any of them except the Arkansas troops... We find Georgia troops in our front; and our [Texan] boys tell them that if they run we will shoot them, and no mistake, and as soon as they find that

the Texans are in their rear, they believe we will shoot
them sure enough..."[2]

So, if you're a Confederate general late in the war running out of
bodies to fill the lines, it might occur to you that there's an obvious
source of new manpower that the Confederacy isn't using: the young
men among those 3.5 million slaves.

Of course, the Confederacy no longer controlled many of them.
Hundreds of thousands of former slaves had fled to Union-controlled
territory or gone into hiding in forests or swamps, so that by 1864
the Confederacy held about half of its original enslaved population.
But it was still a big population, at least 1.5 million people, probably
containing something like "150,000 men of military age." Quite a
pool of future troops, you'd think.

But the Confederate elite absolutely refused to think any such
thing. Arming the slaves was taboo in elite Confederate circles. As
often in the Civil War, it was a foreigner, an immigrant, who insisted
on breaking the taboo and saying that the Confederacy needed to free
and enlist the enslaved.

The foreigner who dared to say this was General Patrick Cleburne,
an odd figure, the son of a Protestant Irish insurgent who was exiled
after the failure of the 1798 United Irishmen rebellion. He grew up
in Arkansas, one of the wildest, most violent parts of the Confeder-

---

2. Foster, Samuel T. *One of Cleburne's Command : the Civil War
   Reminiscences and Diary of Capt. Samuel T. Foster Granbury's
   Texas Brigade, CSA.* p.82 Library of America; First Edition.
   1980.

acy, and had to deal with some local prejudices himself. He and his friend Thomas Hindman got into an argument with some Arkansas Know-Nothings who didn't like Cleburne's accent or attitude. One of the Know-Nothings made his point in the standard manner by shooting Cleburne in the back. Cleburne turned around and shot one of the Know-Nothings. Cleburne survived; the Know-Nothing didn't. That is how discourse, as the grad students call it, worked in antebellum Arkansas: point, counterpoint.

Even after experiencing this kind of xenophobia (which was hardly unique to Arkansas or the south), Cleburne kept his loyalty to his state, and thus to the Confederacy. He joined the CSA with his friend Thomas Hindman, and turned out to be a very good general (unlike Hindman, who was a disaster).

Cleburne should have been given the command of the Army of Tennessee when Jefferson Davis fired Joe Johnston in July 1864. Instead, Davis chose J. B. Hood, who proceeded to destroy that army in a few months of suicidal charges against Union fortifications. Hood's destruction of the Confederacy's second-biggest army was a big factor in ending the war. If Cleburne had been in command, things would have been very different.

Cleburne was, without doubt, a far better general than Hood, but he was never considered for overall command of the Army of Tennessee. That was because he violated the taboo in a January 1864 memo about arming and freeing the slaves, which he read aloud to a meeting of fellow officers.

If Cleburne had sent his memo to President Davis, it could have been suppressed, and he might—might—have kept his reputation. But by reading it to the elite of the western Confederacy, he broke the biggest taboo in his adopted culture. Even so, he was so admired by his colleagues that his letter was co-signed by 13 officers in his division.

Some very high-ranking officers like Hardee even sympathized in private. But they weren't going to take the risk of agreeing with Cleburne publicly.

Cleburne, who saw the war more clearly than US-born colleagues, never understood the horror his proposal caused among Confederate colleagues. In this, he's a lot like that other foreigner, Count Gurowski, who said that the enslaved would "make an excellent peasantry" once freed and given land. Gurowski, like Cleburne, was familiar with European prejudice, which was mostly based on religion and language, but couldn't grasp the intensity of color bias in his adopted country.

Cleburne's letter is a remarkable document. It's a kind of autopsy-in-advance of the Confederacy's cause of death, full of forbidden truths. First, that the Confederacy's current situation is hopeless:

> "...our struggles and sacrifices have...left us nothing but long lists of dead and mangled...Our soldiers...are sinking into a fatal apathy, growing weary of hardships and slaughters which promise no results...If this state continues much longer, we must be subjugated."

Every Confederate with any sense knew this by January 1864. Vicksburg had fallen, the trans-Mississippi was cut off, Grant was in charge, and McClellan—Lee's favorite punching-bag—had been sidelined. Lee's attempted invasion of Pennsylvania had ended in bloody disaster at Gettysburg; France and Britain had made it clear they would never intervene to support a slave power that was already losing.

And worst of all, the Confederacy just didn't have enough man-power to keep fighting.

Cleburne was blunt about the Confederacy's prospects: "Our sin-gle source of supply [for army manpower] is that portion of our white men fit for duty and not now in the ranks. This "source," the supposed slackers or draft-dodgers, was the big hope of Confederate patriots, who kept insisting that if only these slackers would get back in the ranks, all would be well. Cleburne deflates that idea. First of all, he says, many of those who are avoiding Confederate service are now safe behind Union lines: "Most of the men improperly absent...are now without the Confederate lines..."

And if the Confederacy, already over-mobilized, keeps pushing to enlist skilled workers, it will make everything worse: "remov[ing] from the fields and manufactures most of the skill that direct[s] agricultural and mechanical labor..."

Cleburne dismisses Jeff Davis's pet notion of conscripting the very young and very old. He speaks from experience when dismissing these demographics: "...striplings and men above conscript age break down and swell the sick lists more than they do the ranks." That was simple truth, but a truth that diehards like Davis would not accept.

Then Cleburne drops the big one: Slavery, the cause for which the Confederacy fought, has gone, "...from being one of our chief sources of strength at the commencement of the war...[to] one of our chief sources of weakness." This is one of those simple, obvious facts that no one dared mention in Confederate circles. Digression: Beware of anyone who uses the word "obvious" as a pejorative. They generally mean "That's so obvious that SHUT UP ABOUT IT!"

At the beginning of the war, the Confederate elite imagined that they could put every white male into combat because the slaves would do the civilian work. During the first year of the war that seemed possible because Union generals sympathetic to Southern slave owners initially helped them hold on to their "slave property".

Not coincidentally, this first year was the Confederate's most successful. Once Union leaders like Buell and McClellan were replaced by hard-war advocates, slavery was revealed as the "chief source of weakness" for the Confederacy. This was for reasons Cleburne lays out bluntly:

> "Wherever slavery is once disturbed...even by a [Union] cavalry raid, the whites can no longer...sympathize with our cause. The fear of their slaves is continually haunting them...and many of them soon learn to wish the war stopped on any terms."

This is the fragility of slavery, which no one in the Confederacy anticipated. Because they fear their slaves, planters can't coerce them anymore. And without coercion, slaves have become "comparatively valueless to us for labor, and of increasing worth to the enemy" as spies:

> "[Slavery] is an omnipresent spy system [for the Union armies], pointing out our valuable men to the enemy, revealing our positions, purposes, and resources, and yet acting so safely and secretly that there is no means to guard against it."

This was simple fact, shown in the journals of every Union officer. John Beatty describes a dramatic meeting with one enslaved informer:

"A negro came in...to report that the [Confederate] pickets were at his master's house....He was a field hand, bare-footed, horny-handed, and very black, but he knew all about 'de mountings [mountains]; dey can't kotch him nohow. If de sesesh am at Massa Bob's when I git back, I come to-night an' tell yer all.' With these words, this poor proprietor of a dilapidated pair of pants and shirt, started over the mountains. What are his thoughts about the war, and its probable effects on his own fortunes, as he trudges along over the hills? Is it the desire for freedom, or the dislike for his overseer, that prompts him to run five miles of a Sunday to give this information? Possibly both." *(The Citizen-Soldier. "August 1862")*

This field hand's willingness to spend his Sunday, his only free time, running five miles to give the nearest Union forces information on Confederate activity, was a common event as Federal armies moved south—so common, in fact, that many Union officers took heroism like this for granted, or worse.

During Sherman's Georgia campaign, one of his most right-wing generals, Jefferson C. Davis, destroyed a bridge his force had just crossed, abandoning hundreds of the enslaved who were following him to be recaptured and whipped, if not killed. (JCD had already killed a fellow Union officer, "Bull" Nelson, so he had a rep.)

It's striking that Cleburne the Confederate general had a much better understanding of the value of the enslaved to the Union armies than many senior Union officers did. But, as noted in previous chapters, many—not all—of the West Pointers were in open sympathy with the slaveowners and made their contempt for the enslaved, and in fact all black people, very clear. (The quotes are unprintable by current standards.)

Cleburne, the foreigner, saw it differently. Every ex-slave who fled the farms weakened the Confederate supply chain, raised fears among whites left in the "hollow" interior of the over-mobilized Confederacy, and above all, could become an enemy soldier.

Many white officers on both sides affected to sneer at the idea of freedmen as soldiers. Cleburne did not. In fact, he said bluntly—so bluntly that his words might have offended many conservative Union officers, as they did his Confederate colleagues—that black soldiers had proved they could fight: "...the experience of this war has been so far that half-trained negroes have fought as bravely as many other half-trained Yankees."

That assertion, by itself, would have been enough to fry other Confederate officers' brains. They didn't attempt to refute it; they simply felt, as Cobb said, that if it was true, then "our whole theory of slavery is wrong," and that was too much to consider.[3]

What made it even worse was that Cleburne wanted to make ex-slaves into something like full citizens of the Confederacy, guaranteeing them that neither they nor their families could be ripped apart on the auction block:

> "...we must immediately make his marriage and parental relations sacred in the eyes of the law and forbid their [the ex-enslaveds' relatives] sale."

---

3. See: Letter from Howell Cobb to James A. Seddon (January 8, 1865)

Cleburne bursts yet another late-Confederate delusion by pointing out what everybody knew, though no one would say it: Britain and France would never side with the Confederacy while slavery lasted:

> "[Britain and France] cannot assist us without helping slavery, and to do this would be in conflict with their policy for the last quarter century."

But if the enslaved are freed by the Confederacy, "the sympathy of these and other nations will accord with our own, and we can expect from them both moral support and material aid."

All these reasons to arm the slave seem clear, but they provoked something like a Doctor Evil response: "You just don't get it."[4] And Cleburne demonstrates that he *really* doesn't get it with this comment:

> "It is said that slavery is all we are fighting for, and if we give it up we give up all. Even if this were true, which we deny, slavery is not all our enemies are fighting for. It is merely the pretense to establish sectional [Northern] superiority and a more centralized form of government..."

Cleburne's confusion shows through clearly in this passage. Slavery was, in fact, what his colleagues were fighting for. By imposing his Britain vs Ireland model on the USCW, he misapprehends the whole

---

4. *Austin Powers: International Man of Mystery* (1997).

conflict. His concept of the aftermath of a Union victory has much more in common with Ireland after the defeat of the 1798 Rebellion than with the kid-glove treatment given to the Confederates after Appomattox:

> "Every man should endeavor to understand the meaning of subjugation before it is too late...It means that the history of this heroic struggle will be written by the enemy; that our youth will be trained by Northern school teachers; will learn...their version of the war; will be [trained] to regard our gallant dead as traitors, our maimed veterans as fit objects for derision...The conqueror's policy is to divide the conquered into factions and stir up animosity between them, and in training an army of negroes the North no doubt holds this thought in perspective."

If only! Cleburne never imagined that the Northern consensus, the broad centrist consensus, would restore the Confederate states, then to step back and allow the surviving Confederate elite to terrorize the freedmen into something resembling slavery without public auctions. But TBF, most of the Confederate elite shared the perfectly rational view that they—some of them, at least—would be hanged, and their states reorganized as new territories, if they lost the war.

But maybe they did know that they would survive the Northern victors' sloppy, lazy postwar occupation. What makes me wonder about this is a sentence from Cleburne's memo, a sentence that makes me wish we could have a time-traveling drone camera scanning the faces of his audience. Cleburne says bluntly that it's a simple choice between freeing/enlisting the enslaved, and losing the war, then adds:

> "As between the loss of [Confederate] independence and the loss of slavery, we assume every [Confederate] patriot will freely give up the latter—give up the negro slave rather than be a slave himself."

That's the crux of it right there. And all you can say is, "Well, Cleburne, you assumed wrong."

The monetary value of slave "property" was rising almost vertically before 1860.[5] That was why planters would rather risk their sons' lives than "loan" their slaves to Confederate commanders, as nearly all Confederate officers complained in their war journals.

And perhaps there was another reason. Cleburne was an outsider in many ways—not only as an immigrant, but as someone from a remote part of the Confederacy, Arkansas, with no sense of how the U.S. elite operated.

Many of those in his audience, starting with Jefferson Davis, had served in Congress, negotiated with their northern counterparts, experienced many northerners' deep sympathy for the white South. Perhaps those people, for all their apocalyptic wailing as the war turned

---

5. "Figure 2: Average Price of a Slave Over Time Current Dollars" from "Measuring Slavery in 2020 Dollars" by Samuel H. Williamson   and   Louis   P.   Cain https://www.measuringworth.com/slavery.php#:~:text=Using %20these%20measures%2C%20the%20value,for%20the%20enti re%20antebellum%20period

against them, had begun to calculate their chances in a post-defeat USA, and had realized how easy to handle, how lazy and profit-driven, the northern elite was; how deeply it shared their own system of racial hierarchy; and how they could exploit it to reassert their primacy over the post-war South. And felt that, even after defeat, they could re-impose something very close to slavery on their penniless, landless ex-slaves.

You have to wonder if Cleburne had time to realize this before his death.

He didn't have long to puzzle it out. Jefferson Davis outlawed any mention—not just discussion but mention of Cleburne's proposal, saying that any "promulgation" of it would merely cause "discouragement, distraction, and dissension"—which phrase, BTW, shows clearly that the use of three alliterated pejoratives was an old reactionary trick long before William Safire started using it in the Spiro Agnew speeches he wrote.

Davis then refused to consider Cleburne for the command of the Army of Tennessee when he decided to fire Joe Johnston.

Hood, who got the job, was a pure product of Planter culture, without the mental capacity to entertain dangerous notions like Cleburne's. Hood proceeded to destroy his army within a few months—and at Franklin, Tennessee on November 30, 1864, just 11 months after Cleburne stood up and read his memo to a crowd of Confederate officers, Hood finished the job by ordering a suicidal charge across a mile and a half of open ground (half again as long as the fields Pickett's division crossed at Gettysburg) into a fortified Union position.

Cleburne's troops were at the center of the charge. They were slaughtered, and Cleburne died leading them, as he must have realized he would. In a speech before the battle, he asserted the analogy between a post-defeat south and post-1798 Ireland. A few minutes afterwards, he and 7,000 others were dead.

There's a price for getting your historical analogies wrong. Cleburne had seen a solution to the Confederacy's dilemma: arm and free the slaves. And he had seen how that idea was squelched. Most people don't learn because they'd rather die than learn the horrible truths that determine their fate.

Chapter Ten

Turchin's Trial: The Shift To Hard War

*B*eatty's *drunken colonel wasn't the only Union officer to be culled in the chaos of the early years of the war. A much bigger casualty of the policy shift to a hard war was Don Carlos Buell, one of the more extreme advocates of soft war toward the South.*

There was a song Union soldiers sang to taunt Confederates in the Western theater of the Civil War, called "Mister, Here's Your Mule" about stealing a Southern farmer's mule. Every verse describes soldiers playing keep-away with the livestock, but the last line of the last verse changes to "...and Turchin's got your mule!"

Ivan Turchin was the highest-ranking Russian immigrant in the Union Army. A graduate of St. Petersburg military academy, he came to the U.S. in 1856 and was a railroad executive in the Midwest when the war started.

It's remarkable how many Civil War officers were working as railroad executives before the war—a very dynamic profession with a lot in common with commanding troops, since it was basically a job in engineering and logistics. The railroad Turchin worked for was based in Illinois, so he volunteered for the Union. Does this mean he had clear convictions about slavery or secession? Maybe, but he may simply have felt that Illinois, his new home, needed trained officers.

Many immigrants simply "caught the enthusiasm of their neighbors" and joined up with them. This resulted in some ideological oddities, like the strange, doomed career of the Irish immigrant Patrick Cleburne, whose career we saw in the previous chapter.

Cleburne's notions about the Confederate cause show up in the speech he made to his troops before leading them in a doomed charge at Franklin:

> "In Ireland the downtrodden masses had suffered from oppression, yet if the North prevailed the South's condition would be much worse."[1]

That, of course, was ridiculous. The defeated Confederates were treated with great generosity, unlike (very unlike) the defeated Irish. But Cleburne probably really believed his absurd analogy. Things look very different when you're on the ground, young and dumb, imitating all the locals you admire and look up to.

Immigrants to America circa 1860 had to decide fast how they were going to jump, and it wasn't always an easy or obvious choice, though

---

1. *The Confederacy's Last Hurrah: Spring Hill, Franklinn, and Nashville* by Wiley Sword (Open Road, 1993) p.57.

it might seem so now. One group stands out for the clarity of its vision: the Germans and other Central Europeans who'd fled the failed 1848 rebellions. They did understand the justice of the Union cause and the rottenness of the Confederate one.

But Turchin had not come to America as a fleeing radical, like Gurowski and Willich. He was just trying to make a living. In fact, Turchin had been on the oppressor's side of the 1848 Hungarian Revolt, the same one that the West Point cadet E.A. Poe wanted to join.

European wars of the Civil War era were merciless. Soldiers tended to follow the few rules their officers enforced. When they encountered resistance, though, all bets were off and massacres were frequent. For example, I looked up the Spanish city of Badajos simply because it's mentioned in a comic poem I was reading called "Faithless Nelly Gray": *O Nelly Gray! O Nelly Gray!/ For all your jeering speeches,/ At duty's call I left my legs / In Badajos' breaches.* The speaker, Ben Battle, lost his legs at Badajos because Wellington's army was held up there in a long, bitter siege. And, as I discovered, they took revenge on the town by the traditional method: massacre, rape, and pillage. I never heard of the massacre at Badajos and was simply looking up the reference in a comic poem, but this turns out to be one of many massacres in the supposedly clean European wars of the early nineteenth century. Between 200 and 300 Spanish civilians within the walls of Badajoz were killed or injured.[2]

---

2. Gilbert, Adrian, et al. "Siege of Badajoz (1812)", Encyclopedia Britannica. https://www.britannica.com/event/Siege-of-Badajoz-1812Accessed March 12, 2025.

Turchin, whose European military experience was in the very rough school of 1848 counterinsurgency, had the bad luck to serve in the Army of the Ohio under Don Carlos Buell, often called the McClellan of the West. Buell was a strange and unpleasant man who was either an outright traitor, or maniacally devoted to a naïve West Point doctrine of soft war. He was fanatical about protecting the rights and the property of white Southern Confederates. He refused to allow escaped slaves to be sheltered by Union forces, brutally punished any soldier caught stealing produce or livestock from Southern farms, and, as an officer in the regular army, had open contempt for his volunteer soldiers.

But there was another reason for his policy of taking it easy on the South. Buell was a "conservative Democrat," an ambivalent faction for which Civil War historians make constant, unconvincing excuses. Radical Republican officers were not shy about calling these people traitors. If we studied them as we would officers in any other nation's army, we wouldn't hesitate to say that they were more sympathetic to the ostensible enemy, the wealthy whites who ran the Confederacy, than to abolitionists or the enslaved.

Buell owned slaves, "probably voted for James Buchanan" as a sympathetic biographer admits, and married the widow of Richard Barnes Mason, "an aristocratic Virginian...[who] possessed all the peculiarities of a Southerner, accentuated."

Margaret Buell "had been extremely fond of the Mason family and in particular Senator James M. Mason, whom she considered a brother." I talked about James Murray Mason in chapter five, specifically John Logan's revelation that Mason had been scheming with Britain to destroy the Union as early as 1841. Many of Buell's subordinates found his sympathy for the Southern elite and scorn for his own soldiers bizarre. Colonel John Beatty described it as the "dancing-master policy" toward the South:

> "Buell...is inaugurating the dancing-master policy: 'By your leave, my dear sir, we will have a fight; that is, if you are sufficiently fortified; no hurry; take your own time.' To the bushwhacker: 'Am sorry you gentlemen fire at our trains from behind stumps, logs and ditches. Had you not better cease this sort of warfare? No do, my good fellows, stop, I beg of you.' To the citizen rebel: 'You are a chivalrous people; you have been aggravated by the abolitionists into subscribing cotton to the Southern Confederacy; you had, of course, a right to dispose of your own property to suit yourselves, but we prefer that you would, in future, make no more subscriptions of that kind, and in the meantime we propose to protect your property and guard your negroes." *(The Citizen-Soldier "July 1862")*

Beatty poses his frustration with Buell's softness in an antithesis with Turchin's alleged severity, rejecting both:

> "Turchin's policy is bad enough; it may indeed be the policy of the devil; but Buell's policy is that of the amiable idiot." *(July 1862)*

Beatty is actually being generous here, since no one else ever called Buell amiable. Like most West Point mediocrities, Buell was ye compleat martinet, who saw no reason to treat volunteer troops any better than regulars who were to be beaten into obedience.

The reason Beatty was musing on Turchin's conduct was because he was one of the officers assigned in July 1862 to investigate Turchin for "outrages" against Confederate civilians. Beatty introduces his colleagues on the panel, which included future president James Garfield, and says:

> "The first case to be tried is that of Colonel J.B
> . Turchin, 19th Illinois. He is charged with permitting
> his command, the eighth brigade, to steal, rob, and
> commit all manner of outrages." *(July 1862)*

Beatty's ambivalence about prosecuting Turchin is clear in his description of the charges:

> "There are many wealthy planters in this section.
> One...swears that Turchin's brigade robbed him
> of twelve hundred dollars' worth of silver plate.
> Turchin's brigade has stolen a hundred thousand dol-
> lars' worth of watches, plate and jewelery, in Northern
> Alabama. Turchin has gone to one extreme, for war
> can not justify the gutting of private houses and the
> robbery of peaceable citizens..." *(July 1862)*

But, as Beatty makes clear, he has no more sympathy for Buell's "amiable idiot" softness. After a year of getting shot at by these supposed peaceful citizens and rejecting pleas for sanctuary from slaves who were the Union soldiers' only real friends, Beatty and his fellow officers were beginning to loathe Buell's policies:

> "[At] every plantation negroes come flocking to the roadside to see us. They are the only friends we find. They have heard of the abolition army, the music, the banners, the glittering arms; possibly they hope that their masters will be humbled and their condition improved, gladdens their hearts and leads them to welcome us with extravagant manifestations of joy."
> *(March, 1862)*

Like many Union officers, Beatty started the war assuming that he would find peaceful citizens among Southern whites. He hoped to demonstrate that the Union army was not an "abolitionist army." In June 1861, marching into the Southern heartland, he wrote:

> "The country people here have been grossly deceived by their political leaders. They have been made to believe that Lincoln was elected for the sole purpose of liberating the negro..."

By March 1862, after noting the shock his troops felt at seeing signs like "NEGROES BOUGHT AND SOLD" in Kentucky, Beatty and many other junior officers were more than willing to become an "abolition army" in reality:

> "[General Buell] issued [an] order requiring us to surrender the negroes to claimants and to 'keep colored

folks out of our camp hereafter'...I obeyed the order
promptly; commanded all the colored men in camp
to assemble at a certain hour to be turned over to their
masters but...the scamps, I fear, took advantage of my
notice and hid away, much to the chagrin of all who
desire to preserve the Union as it was."

This phrase "the Union as it was" had become the slogan of
pro-slavery Democrats like McClellan and Buell. It meant, simply, the
Union with slavery.

Beatty's experience in the first months of 1862, as Buell's army
moved slowly in its dancing-master way, southward, radicalized him
and many other soldiers. The "peaceable citizens" they hoped to find
simply didn't exist. They met very few of Lincoln's beloved Southern
Unionists.

For Beatty, the last straw was an incident in May 1862 after his
troops "took the cars [railroad] for Huntsville" (Alabama):

"At Paint Rock the train was fired upon, and six or
eight men wounded. As soon as it could be done, I
had the train stopped, and, taking a file of soldiers,
returned to the village....Calling the citizens together,
I said to them that this bushwhacking must cease. The
federal troops had tolerated it already too long. Hereafter, every time a train was fired upon we should hang
a man; and we would continue to do this until every
house was burned and every man hanged between
Decatur and Bridgeport....I then set fire to the town,

took three citizens with me, returned to the train, and proceeded to Huntsville." *(May, 1862)*

Beatty's reprisal was pure insubordination, violating the letter and spirit of Buell's policy. But by this point Buell was already in trouble for his reluctance to take on Braxton Bragg. Buell would be replaced in October 1862 when the need for a hard war got through to the thick skulls in Washington. But the reaction to Beatty's reprisal at Paint Rock shows that Buell had already lost the allegiance of his brigade-level officers:

> "General Mitchel [Beatty's immediate superior] is well pleased with my action in the Paint Rock matter. The burning of the town has created a sensation, and is spoken of approvingly by the officers and enthusiastically by the men." *(May, 1862)*

General Ormsby Mitchel was in overall command of the push southward into Alabama after being left in charge at Nashville when Buell moved west to join Grant at Shiloh. Mitchel, who Beatty says disapproved of Buell's pro-Southern policies, was an aggressive commander, one of the few holding high rank in the Union Army at the time. In early May 1862 Mitchel's force took Huntsville, a major Alabama city. This was a shock to the Confederate supporters who were used to Buell's slow velvet-glove advances. Their response was impromptu guerrilla war by mixed groups of armed civilians and the dreaded "irregular cavalry."

This term, irregular cavalry, became a pejorative in the course of the war but, typically, Buell insisted on considering them legitimate combatants, though they fought without uniforms and largely by sudden ambushes, commonly called bushwhacking.

One of Turchin's regiments pushing south into Athens, due west of Huntsville, was attacked in roughly the same way that Beatty's force was attacked in Paint Rock, to the east. They had been harassed by the same bushwhackers who'd attacked Beatty's train and were in a foul mood, especially because Buell's standing orders forbade them from carrying out any reprisals.

On May 2nd, after repelling the bushwhackers (or, if we want to be polite, irregular cavalry) that had surprised them in Athens, Turchin's troops committed what became known in Lost Cause historiography as "the Sack of Athens."

You can't help chuckling at the presumption of that term. We're not talking about the Persians storming the Acropolis here. That's the problem with provincial 19th-century America (well, one of the problems anyway): every time people built three or four cabins at a crossroads they named it Athens or Philopolis and imagined it as the future seat of wisdom.

At any rate, something happened in Athens when Turchin's troops seized the town on May 2. Exactly what isn't clear but there were certainly departures from the hands-off-we're-your-friends policy. And at this point in the war, when Confederates were convinced they could do pretty much anything without provoking any reprisal from Union forces, it was a shock. And not just a shock; rather like Sherman's March two years later, it was a bitter humiliation for white Southerners, never to be forgotten. The news that Union troops had reacted at all to provocation also enraged Buell and many old regulars among the officer corps though no one even claimed that Union troops had burned buildings or assaulted citizens.

Turchin's foreign background, particularly the fact that he was a Russian, contributed to the hysteria. As Walter Sobchak, distinguished military historian, would say, "Nothin' ever changes."

James Garfield, who was one of the seven officers empaneled to court-martial Turchin, attributed the "sack" to Turchin's training in the ruthless Tsarist army. The legend grew that Turchin had said, "For three hours I turn my back." It sounds more like Timur than a Union commander but, again, it's very hard to get any solid account of what happened that day. Most accounts, in fact, lean heavily toward the Lost Cause narrative of outrage against defenseless Southerners. This happens again and again with Lost Cause narratives: though there isn't much evidence for their claims, they win out simply because their resentful adherents care more than the Union side, which lost interest in the South soon after winning the war. Resentment is a wonderful preservative.

What does seem likely is that Turchin's men (practical Yankees that they were) looted the houses of the rich in Athens. Beatty clearly believed this version. And in the spring of 1862 that was cause enough for a court martial. Two years later, it would hardly have been noticed (I wonder if there's any account of how many soldiers in Sherman's March went home and built Victorian McMansions with unexplained funds).

Beatty's war journal gives a detailed account of his time on the court martial panel, but most of it is about the fun he had getting a few weeks off from managing troops in the field. There was many a drunken lunch, a lot of typical practical-jokery (they were big on practical jokes, these Union officers), but very little about the issues in the trial.

The impression you get is that none of the seven members of the court martial were especially shocked by Turchin's alleged misconduct. From Mitchel on down, the subordinate commanders were heartily sick of Buell's open sympathy for the enemy and brutality toward his own men. Some radicals, like Carl Schurz, were openly calling Buell a Southern sympathizer by this time. Buell's post-war conduct and correspondence support that view.

During the trial, Beatty confided in his diary that he supported policies that sound very much like the "outrages" of which Turchin is accused:

> "If we make the country through which we pass furnish supplies to our army, the inhabitants will have less to furnish our enemies. The surplus products of the country should be gathered into the Federal granaries, so that they could not, by possibility, go to feed the rebels." *(July, 1862)*

He and his colleagues had clearly lost patience with the commonplace that Unionists and Confederates were simply fellow Americans who differed in their opinions: "The opinions from which we differ in this instance are treasonable."

So the court martial was a failure from Buell's point of view and simply confusing for everyone else. On July 19th, one day after Beatty wrote the preceding passage about an end to coddling Confederates, he notes, "Turchin has been made a Brigadier." Two days later, he noted without comment, "Colonel Turchin's case is still before us. No official notice of his promotion has been communicated to the court."

Turchin was promoted before there was any chance to condemn him thanks to two factors: his effectiveness as a commander, and the lobbying efforts of his extraordinary wife Nadyezhda Lvova Turchin.

As the Chicago Times wrote, "Truly, in the lottery matrimonial Col. Turchin had the good fortune to draft an invaluable prize."

Mme Turchin deserves a book, in her own right, but for present purposes her lobbying on her husband's behalf in Chicago and Washington was successful not just because of her own persuasiveness and charm (which were apparently remarkable) but because Lincoln's administration, particularly Secretary of War Stanton, was sick of the bizarre pro-Southern-white policies of the West Pointers.

Buell was being commonly disparaged as "the McClellan of the West" and would be ousted from command, replaced by Rosecrans, within a couple of months of Turchin's effective acquittal. The technical reason for letting Turchin's charges drop was that after his promotion he outranked six out of seven members of Beatty's court-martial team. In reality, as Beatty's account makes clear, the seven buddies were having a great time and had no interest in punishing one of their more effective colleagues simply to further the ridiculous "dancing-master policy" of the hated Buell.

The era of the soft war was over and Turchin had the mule.

Chapter Eleven

Why Sherman Was Right to Burn Atlanta

A mericans still argue about the Civil War, and when they do the focus always turns to Sherman and his scorched-earth campaign across the South. Part of the problem is that many of the Americans who still care about the war have learned old grudges in their childhood. It would help if more Americans were exposed to the world standard of ambient violence in wartime, but that's not likely to happen with the American education system so in the meantime, here's why I think Sherman's firebug habits were remarkably effective.

Sherman's March to the Sea was the most shocking and effective campaign of the Civil War. He led his army out of Atlanta on November 15, 1864, and marched them across rural Georgia, which had been largely untouched by the war, burning as he went. The burning of Planters' mansions and public buildings would not have shocked

Europeans, but it stunned the Confederate public. To this day, there are descendants of Confederate families who regard Sherman as a war criminal and his March to the Sea as one of the most outrageous war crimes in history.

They should have known because Sherman started the campaign by burning his base, the city of Atlanta. This outraged but did not teach any Confederates; their indignation (which seems undiminished more than a century after the fact) was expressed in a guest editorial in the New York Times by one Phil Leigh on the 150th anniversary of the start of the Atlanta Campaign.[1]

Leigh poses a simple question "Who burned Atlanta?" and comes up with a simpler, moralizing answer: "Sherman, that bad, bad man!"

It's a strange question and a stranger answer. The United States military has bombed, burnt, depopulated, defoliated, and otherwise damaged many cities in many countries, but there seems to be an unspoken rule that "We" don't do that to "Each Other." This was not Sherman's attitude. He had seen the soft-war approach tried by commanders like Buell and McClellan in the early years of the war and he, along with their other subordinates, had seen it fail dramatically. As far as he was concerned, war was famously Hell. That quote got remembered but what seems to surprise an infinite series of Confederate generations is that he actually meant it.

In fact, Sherman didn't inflict anything like Hell on Georgia. Hell would have been what Tsarist armies inflicted on Poland after they crushed the 1830 revolution or British armies inflicted on rebels in dozens of their colonies. Hell would have involved wholesale massacres

---

1. "Who Burned Atlanta?" by Phil Leigh. The New York Times. November 13, 2014. p.148

of civilian populations. That never happened. Sherman is not even accused of doing that. What he did was march across Georgia looting and burning, and he did it after long experience in the futility of a soft-war strategy that allowed Confederate civilians to go about their business while their relatives sniped at occupying Union forces. That simply didn't work, and Sherman felt that since "the South is hollow, all hollow inside," the most effective way he could teach the population of the Confederate states that they had lost was to show them their defeat in banners of black smoke spreading across the skies of Georgia.

Sherman had tried, before the war, to slap the white Southerners awake. He was no abolitionist or radical Republican. In fact, he's notably unsympathetic to the enslaved, alas, but he had lived and worked in the South before the war and had seen, in his clear-headed way, what the South could and could not do in military terms.

Speaking to Southern friends before the war, Sherman tried to show them in the bluntest terms why they had no chance:

"You people of the South don't know what you are doing. This country will be drenched in blood, and God only knows how it will end. It is all folly, madness, a crime against civilization! You people speak so lightly of war; you don't know what you're talking about. War is a terrible thing! You mistake, too, the people of the North. They are a peaceable people but an earnest people, and they will fight, too. They are not going to let this country be destroyed without a mighty effort to save it... Besides, where are your men and appliances of war to contend against them? The North can make a steam engine, locomotive, or railway car; hardly a yard of cloth or pair of shoes can you make. You are rushing into war with one of the most powerful, ingeniously mechanical, and determined people on Earth —right at your doors. You are bound to fail. Only in your spirit and determination are you prepared for war. In all else you are totally unprepared, with a bad cause to start with. At first you will make headway, but as your limited resources begin to fail, shut out from the markets of Europe as you will be, your cause will begin to wane. If your people will but stop and think, they must see in the end that you will surely fail."[2]

That was Sherman's advice to the South before the war even began. And he was, as usual, absolutely right. But he was talking like a grown-up to people who didn't want to think like adults. Their whole society was based on horrible lies—"a bad cause to start with"—which happened to have made them a lot of money while imbuing them with a deep aversion to any challenge to the moral legitimacy of their dominion over the enslaved. So they stuffed themselves, as Mark Twain said, with copious doses of the worst "chivalrous" nonsense they could find, like Walter Scott's pseudo-medieval novels, and went

---

2. Lewis, Lloyd (1993) [1932]. *Sherman: Fighting Prophet*. University of Nebraska Press. p. 138; Exchange between W. T. Sherman and Prof. David F. Boyd, December 24, 1860, attributed to "Boyd (D.F), mss. [manuscripts] in possession of Walter L. Fleming, Nashville, Tenn.".

off to cause the biggest slaughter of their fellow Americans in history, a body-count far higher than the sum total of all Americans killed in all wars with other countries.

Oh, but that was glorious, for idiots like Phil Leigh. What was non-glorious was Sherman burning Atlanta. You see what Sherman was up against? That's why his campaigns, unlike any other Union general's and in fact any other waged by an American commander until the age of "hearts and minds" warfare dawned a century later, were designed, above all, to smack awake a crazed and homicidally delusional population. Like John Wayne slapping some hysterical private, Sherman tried, in everything he said and did, to make the South face reality.

Sherman knew the wider world, and tried to warn the arrogant provincials who ran the Confederacy what it meant to them—all the peoples wiped out of existence for far less sustained craziness than the South was demonstrating, and all the eager immigrants waiting to take the traitors' places. In a letter to a fellow Union officer written near the beginning of his March to the Sea, Sherman reviews the narrowing options that the South refuses to face when it continued the war after the fall of Vicksburg (July 4, 1863) and the Confederate defeat at Gettysburg (July 1-3, 1863) had made it clear that the South would not win the war. There was no strategic justification for continuing the war after these two massive defeats, but by the beginning of 1864, it was clear that the South would fight on with no hope. In a letter to his adjutant R.M. Sawyer, Sherman laid out the consequences of the Confederacy's repeated refusal to face its strategic defeat:

> "If [the Confederates] want eternal war, well and
> good; we accept the issue, and will dispossess them
> and put our friends in their place. I know thousands

and millions of good people who at simple notice would come to North Alabama and accept the elegant houses and plantations there. If the people of Huntsville think different, let them persist in war three years longer, and then they will not be consulted. Three years ago by a little reflection and patience they could have had a hundred years of peace and prosperity, but they preferred war; very well. Last year they could have saved their slaves, but now it is too late. All the powers of earth cannot restore to them their slaves, any more than their dead grandfathers. Next year their lands will be taken, for in war we can take them, and rightfully, too, and in another year they may beg in vain for their lives. A people who will persevere in war beyond a certain limit ought to know the consequences. Many, many peoples with less pertinacity have been wiped out of national existence."[3]

What Sherman's march did was to make a public demonstration of the impotence of the South's aristocratic males. He said so himself: "My aim...was to whip the rebels, to humble their pride, to follow

---

3. Letter to R.M. Sawyer (January 1864)

them to their inmost recesses, and make them fear and dread us. 'Fear of the Lord is the beginning of wisdom.'"[4]

The taking and burning of Atlanta were just one more chance to slap the South awake, as Sherman saw it. When he was scolded—by people who were in the habit of whipping slaves half to death for trivial lapses—for his severity toward the (white, landowning) people of Atlanta, he replied, in his "Letter to Atlanta," in a way that shows how patiently he kept trying to talk grown-up sense to an insane population:

> "You cannot qualify war in harsher terms than I will. War is cruelty, and you cannot refine it; and those who brought war into our country deserve all the curses and maledictions a people can pour out. I know I had no hand in making this war, and I know I will make more sacrifices to-day than any of you to secure peace. But you cannot have peace and a division of our country...The only way the people of Atlanta can hope once more to live in peace and quiet at home, is to stop the war, which can only be done by admitting that it began in error and is perpetuated in pride.

> "You have heretofore read public sentiment in your newspapers, that live by falsehood and excitement;

---

4. "Northern Soldiers: William T. Sherman" Chapter V in *Patriotic Gore* by Edmund Wilson. (originally published Norton, 1962). Reprinted as an e-book by Farrar, Straus and Giraux.

and the quicker you seek for truth in other quarters, the better. I repeat then that, by the original compact of government, the United States had certain rights in Georgia, which have never been relinquished and never will be; that the South began the war by seizing forts, arsenals, mints, custom-houses, etc., etc., long before Mr. Lincoln was installed, and before the South had one jot or tittle of provocation. I myself have seen in Missouri, Kentucky, Tennessee, and Mississippi, hundreds and thousands of women and children fleeing from your armies and desperadoes, hungry and with bleeding feet...But these comparisons are idle. I want peace, and believe it can only be reached through union and war, and I will ever conduct war with a view to perfect an early success."[5]

Seems clear enough, right? "I just took your city, and out-thought as well as out-fought your generals and troops (and by the way, just to lay another fond Southern myth to rest, the Confederate troops who faced Sherman's army were inferior, not just in numbers or equipment, but man-for-man, one-on-one, as they showed in dozens of battles)—so are you going to wake up and stop whistling Dixie, you loons?"

---

5. Letter to James M. Calhoun, Mayor, E. E. Rawson and S. C. Wells, representing City Council of Atlanta. September 12, 1864.

The answer was obvious: No, they weren't. They still haven't, as Phil Leigh's nasty little commemoration of Sherman's March demonstrates. You can't fix crazy, and it seems to breed true down the generations. Crazy people don't need, or want, evidence. They prefer anecdotes with crying little girls. So here's Phil Leigh's case that burning Atlanta was a bad thing:

> "One Michigan sergeant conceded getting swept up
> in the inflammatory madness, even though he knew it
> was unauthorized: 'As I was about to fire one place a
> little girl about ten years old came to me and said, 'Mr.
> Soldier you would not burn our house would you? If
> you did where would we live?' She looked at me with
> such a pleading look that ... I dropped the torch and
> walked away."

Yes, one Michigan soldier was overcome by sentimentality and "dropped the torch." But that torch, as it were, was passed to stronger hands, and Atlanta burned. As it should have. You know what's worse than a little girl asking "Mister Soldier" not to burn her house? Getting your leg sawed off by a drunken corpsman after a Minie ball fired by traitors turned your femur into bone shards. Or getting a letter that your son died of gangrene in one of those field hospitals where the screaming never stopped, and the stench endured weeks after the army had moved on. Those are the realities of war that Sherman hated—truly hated, which is something you can't say by any means about most successful generals.

Sherman never forgot those horrors. I repeat, he was one of a very few great generals who genuinely hated war, and he never lost a chance to say so:

> "I confess, without shame, that I am sick and tired of fighting—its glory is all moonshine; even success the most brilliant is over dead and mangled bodies, with the anguish and lamentations of distant families, appealing to me for sons, husbands, and fathers ...it is only those who have never heard a shot, never heard the shriek and groans of the wounded and lacerated... that cry aloud for more blood, more vengeance, more desolation."[6]

Sherman never stopped talking like this, even after the war, when memories dimmed and a sentimental nostalgia became the norm among aging Union veterans. In 1880, Sherman made a speech in Columbus, Ohio, to a crowd of veterans:

> "There is many a boy here today who looks on war as all glory, but, boys, it is all hell. You can bear this warning voice to generations yet to come. I look upon war with horror, but if it has to come, I am there."

Here again we see Sherman in his true glory, a cold, bright mind in a crowd of sentimental Victorian killers. I only truly love two Civil

---

6. Letter from May, 1865.

War commanders, Sherman and George Thomas, the best of all of them. But Thomas was a softer man than Sherman, too tender by half to see what Sherman saw. Sherman saw the horror full-on, and never flinched.

But that horror just doesn't register with the Phil Leighs of the world. As far as they're concerned, it was glorious to kill 300,000 loyal American soldiers in defense of the vilest social system since Sparta.

As far as the Times' resident neo-Confederate is concerned, the war was going swimmingly until Sherman came along and bummed their high by abandoning their ersatz chivalry and showing the Planters' sons that they had already lost. They only way he could do that, the only way that would get through to such a rock-headed crowd, was to turn their stately homes into giant billboards composed of black smoke. In the Philip K. Dick novel *Ubik*, a character has to improvise a billboard to tell his former comrades that they are now dead. He uses an advertising jingle to do so in a way that will be memorable enough to reach their impaired consciousness: *Violets are blue, / Roses are red, / I'm the one that's alive, / You're all dead!*

The Confederacy was dead after the twin defeats of Vicksburg and Gettysburg. The Mississippi was closed to the Confederacy, the Atlantic blockade was more effective each month, and—most important—the Confederacy was running out of white males to use as cannon fodder.

But Sherman saw, as few others had, that the craziness of the white South was bone-deep, and could never fully be eradicated. He wouldn't have been surprised to read Phil Leigh's spitball-commemoration of his Atlanta victory. What Sherman did hope—and it was a realistic hope, fulfilled by history—was to suppress the South's craziness for a few generations:

"We can make war so terrible and make [the South]
so sick of war that generations pass away before they
again appeal to it."[7]

And it worked; it wasn't until the twenty-first century that these
neo-Confederates dared to raise their heads and start hissing their crazy
nonsense in neutral venues like The New York Times. So Sherman's
alleged brutality was not a matter of blame, or a regrettable side-effect
of his campaign. It was the point of his campaign. Sherman began with
the goal of humiliating a Southern white elite consumed by delusions
of superiority.

Sherman burned Atlanta for two reasons, both perfectly sound:

1. Because no sane general, planning to send an army of more
   than 60,000 men across the enemy's heartland with no
   supply line or hope of reinforcement, would leave a major rail/supply center like Atlanta intact in his rear. Burning Atlanta was a no-brainer. Any commander would have
   done the same, but very few would have dared undertake the
   march from Atlanta to the Sea at all. It was so radical a plan
   that British military historian B. H. Liddell Hart claimed it
   marked Sherman as "the first modern general" and placed
   him alongside Napoleon and Belisarius as one of the greatest
   commanders of all time.

2. Because every column of smoke rising from a burning mansion, barn, or granary was intended by Sherman as a signal

---

7. Sherman, William T. *Memoirs*, 2nd ed, 2 vols. (New York, 1886),
II, 126 -127. P.126

to a psychotically stubborn, deluded Confederate (white, landowning) population that they had lost, and that every additional life lost was, as he kept trying to tell them, an atrocity, a crime far greater than property destruction.

Sherman never admitted to ordering the burning of Atlanta, because—let's be honest here—there are two rules for American wars: What we do to foreigners, and what we do to other Americans—and for some reason, most historians persist in considering the slave-selling traitors, America-hating swine who ran the Confederacy as Americans. So we could never treat them as we did the people of, say, Tokyo or Dresden, even though the people of those two cities were never responsible for killing so many Americans as the Confederates did.

So Sherman said only this about the burning:

> "Though I never ordered it, and never wished for it,
> I have never shed any tears over the event, because I
> believe that it hastened what we all fought for, the end
> of the war."[8]

He, unlike the Phil Leighs of the world, was thinking about all the horrors of endless guerrilla war: "If the United States submits to a division now, it will not stop, but will go on until we reap the fate of Mexico, which is eternal war..."—which terrified sane grown-ups both North and South, including Robert E. Lee, who told his aides that it

---

8. "Who Burned Columbia" by James Ford Rhodes. The American Historical Review, Vol. 7, No. 3 (Apr., 1902), pp. 485-493 p.489 p.158

was the horror of guerrilla war that made him accept the humiliation of surrender. When the very young, excitable General Porter Alexander proposed that the Army of Northern Virginia literally head for the hills and try guerrilla warfare, Lee answered like a real grown-up:

> "You and I...must consider its effect on the country [i.e. the Confederacy] as a whole. Already it is demoralized by the four years of war. If I took your advice, the men would be without rations and under no control of officers. They would be compelled to rob and steal in order to live. They would become mere bands of marauders, and the enemy's cavalry would pursue them and overrun many sections they may never have occasion to visit. We would bring on a state of affairs it would take the country years to recover from. And, as for myself, you young fellows might go bushwhacking, but the only dignified course for me would be to go to General Grant and surrender myself and take the consequences of my acts."[9]

Lee wasn't as sensible as he could have been because any sane Southern officer knew very well that after the twin defeats at Vicksburg and Gettysburg, the lousy grand old cause was lost and all deaths from now on were completely in vain. But at least he knew that guerrilla war usually inflicts ten casualties on the occupied, i.e. the South, for every one inflicted on the occupier, i.e. the Union troops.

---

9. *R.E. Lee: A Biography* by Douglas Southall Freeman (Charles Scribner's Sons, 1934) p.123

But then Lee had moments of lucidity in an otherwise chivalry-warped consciousness; the Phil Leighs among us have none.

Sherman was, by contrast, the most grimly sane American ever born—and compared to the endless, mindless brutality of guerrilla war—a Jesse & Frank James world, a Quantrill world, metastasized across the continent, compared to which burning a few houses was a wholesome purgative.

Of course, this is all lost on the Phil Leighs of the world, who—for reasons that cut deep into the ideology of the American right wing—always take burnt houses too seriously, and dead people far too lightly. To them, burning a house is a crime, while shooting a Yankee soldier in the eye is just part of war's rich tapestry. So their horror of messing with private property joins their sense of emasculation, and their total ignorance of what war on one's home ground actually means, to form a sediment that could never have been cured, even temporarily, except by the river of armed men Sherman sent pouring south and east from Atlanta on November 15, 1864. That march and its attendant plumes of smoke woke them for a little while, at least—long enough to quicken the end of the war and save thousands of lives.

That was all Sherman hoped for. He'd spent time with these guys, and knew they could never really be cured:

> "...Sons of [Southern] planters, lawyers about towns,
> good billiard players and sportsmen, men who never
> did any work and never will. War suits them..."[10]

---

10. Major-General Henry W. Halleck, September 17, 1863.

Well, they've gained about 60 pounds per capita and forgotten how to ride a horse, but they're still around, still sulking, and, thanks to the New York Times, they've been able to let the rest of us know it. After all, what good is a 150-year sulk if nobody notices it?

Chapter Twelve

Keep the Home Fires Burning: Civil War Arson

S herman wasn't the only arsonist operating in the US in the 1860s Southern agents were totally in favor of arson and in fact tried to burn Manhattan. Incredibly, they failed. In this article, I deal with a very memorable text, the diary of a Southern boy dying of an incurable disease who happens to mention the pattern of arson in his town during the war. It's a feel-good story.

I was reading the world's most depressing book when I noticed something relatively cheery: an awful lot of houses burned in Macon, Georgia, in the Civil War years.

You might say houses burning are not the classic definition of happy-news items, but that's because you don't know what book I was reading: *The War Outside My Window*,[1] which should have probably been titled *Schopenhauer Was Right: We Are in Hell.*

It's the diary of Leroy Gresham, a boy from a wealthy family in Macon who is slowly and painfully dying of spinal tuberculosis. The book ends when Leroy finally dies in March, 1865, a few weeks before Lee surrendered.

For Leroy, life and the war end almost simultaneously. Leroy's war years are spent lying in bed, in constant pain. So his meditations on the war are continually interrupted when the pains in his leg and back become unbearable, as here:

"War! Thou demon that ravishes fair countries, stay thy mad career. Saw off my leg."

"Saw off my leg" has nothing to do with the war. It just means that, in the middle of trying to write a typically high-flown Victorian apostrophe to the war, Leroy's pain becomes unbearable.

I gotta tellya, they say American school curricula are becoming too soft, too eager to shield the kiddies from harsh realities, but I saw an edition of this book with the advertisement, "A Comprehensive Curriculum Guide for Teachers."

So they're letting the kiddies read this book? Well then, get ready for a generation of true nihilists, because even if this book is assigned only for homeschooled neo-Confederate kids, those kids are never going

---

1. *The War Outside My Window: The Civil War Diary of LeRoy Wiley Gresham, 1860-1865 by* Janet Elizabeth Croon (Savas Beatie, 2019)

to be the same. One hour spent reading Leroy's diary will undo a thousand hours of memorizing scripture, or Ayn Rand.

Tuberculosis of the spine means that the spinal column is slowly being consumed by bacteria. In other words, he's being eaten alive. His diary is all about slow death, his own, and almost simultaneously, the Confederacy's. On bad days, Leroy is given morphine or "Dover's powders," an opium-based OTC analgesic. But his parents try to avoid giving him opiates too often, so most entries are stoic notes on whether any of his "abscesses" are ready to burst.

For Leroy, a good day is when an abscess bursts and drains, the pus pours, and he can sleep. But they take a long time to swell and burst, and on those days, Leroy can only lie in bed, relaying news about "the war outside [his] window" to his diary and trying to distract himself with the books, pastries, and pets his family gives him.

The pets are the worst. Leroy's entries are a mix of dying puppies, soldiers' funerals, and his own medical nightmare. Here's a typical entry:

> "My dog Price continues to suck eggs and I have issued orders for his removal. I cut off some of his hair as a memento. Cousin Helen is here on her way to Maryland in search of Capt. Plane's body."

One after another, the gift puppies die—killed by disease, shot by his family for the capital crime of "sucking eggs" from the hen house, or just randomly killed by strangers who like shooting dogs.

And the news from the front quickly turns grim as well. Finally, in an entry written a few days after the climactic moment of the war: the twin defeats of Gettysburg and Vicksburg, Leroy loses another puppy and decides not to accept any more:

> "Thursday July 9, 1863. Thomas shot 'Forrest' [Leroy's latest puppy] today, a sad and cruel act for which I am very sorry. I don't want any more puppies."

It's all like that. So that's why, when Leroy mentions local arsons in the same journal entry, it really is a mood-brightener.

> "On Tuesday night the Brown house was fired and put out. On Wed. morn the Lanier house was fired and after considerable trouble extinguished. Today about 11 am the alarm was again sounded that it was on fire in the garret. The incendiary had fired it with paper saturated with turpentine."

At this point, I started looking for Leroy's mentions of arson in Macon during the war years. There were many:

1861

April 18: "The Granite Hall was burned down last night along with Mr. Bradden's store...Virginia went out [seceded.]"

May 8: "There was a fire yesterday. Mr. Warren's stable back Dr. Fitzgerald."

May 17: "College caught fire but was immediately put out."

May 25: "Cotton Avenue [?] burned last night."

July 1: "There was a fire last night and it burnt from Father's store to the Rio Grande. It was hard work to save them."

November 12: "The Union men of Tennessee have burned several railroad bridges on the East Tennessee & Virginia road."

December 4: "There has been an accident on the Central road; the track was torn up about 20 yards on level ground, 3 cars smashed. Nobody killed. They've hung 2 traitors in East Tennessee, a great wonder. It looks like we never hang anybody that deserves it."

December 13: "There is awful news this morn. Charleston has been visited by a most terrible fire. 1/3 (one third) of the business part of town was utterly destroyed...The fire was started, I have not a doubt, by some 'Southern Yankees.'"

1862

December 12: "Mrs. Stevens was burnt out and lost every rag of clothing and did not save a thing. A negro girl has confessed to having done it."

1863

February 23: "An attempt to set fire to Mr. Lanier's house last night, but was discovered in time to put it out." (There are at least two attempted arsons at Lanier properties mentioned in Gresham's journal.)

March 12: "Mr. Hardin Johnson's house was burned last night and the family barely escaped with their lives. It was set in 3 places..."

March 15: "Mr. Freeman's stable was burned in broad daylight yesterday, and about day [dawn], "The Republic" etc. were burned. 50 bags cotton were destroyed."

July 1: "A dwelling house was burned down last night about 2:30 a.m."

July 9: "On Tuesday night the Brown house was fired and put out. On Wed morn the Lanier house was fired and after considerable trouble extinguished. Today about 10 or 11 am the alarm was again sounded that it was on fire in the garret. The incendiary had fired it with paper saturated with turpentine. After incredible efforts the gallant firemen succeeded in extinguishing it." (Lanier House Hotel)

So it's clear that somebody was lighting a lot of fires in Macon. Fire was always a good weapon for those with no others, and especially powerful in towns made of wood, heated by wood fires, in which cooking was done on wood stoves, and no piped water. New Orleans, the biggest city in the Confederacy, only got piped water in 1869, four years after the war ended.

Macon allegedly had a waterworks as early as 1842. But there's no evidence of it in Leroy's diary. All these fires were fought by bucket brigades, with very little success. So fire was a very effective weapon for insurgents with no weapons but paper soaked in turpentine.

Who were they?

Mary Boykin Chesnut, a member of the Confederate elite, thought she knew very well:

> "Last night a house was set on fire; last week two houses. 'The red cock crows in the barn.' Our troubles thicken, indeed, when treachery comes from that dark quarter."

The phrase "that dark quarter," refers to the enslaved population, the only part of the population in most southern cities which was assumed to be pro-Yankee.

Of course they might not have been the only Union sympathizers. There may have been white Southern arsonists operating in Charleston, Macon, and other cities, as they definitely were in East Tennessee. But most white Unionists lived in hill districts not suitable for cotton planting. Macon was deep inside the cotton belt. The few white Unionists in towns like Macon had been expelled or terrorized into fleeing—or simply killed—in 1861.

So the arsonists probably were those Mary Chesnut suspected, enslaved people striking back one of the few ways they could. Fires bloomed all over the Confederacy:

"...local Unionists had already set fire to 'houses and other buildings' of pro-secession residents." (W North Carolina)

"Suspicious incidents of arson multiplied throughout the south."

"Another servant [of Jeff Davis] tried to torch the presidential mansion."

"Arson struck Lynchburg [VA] enterprises suspected of profiting from the food crisis."

"[cotton] gin-houses, dwelling-houses, and barns, and the court-house of Greene County have been destroyed by [arson] fire...bridges have been burned and ferry-boats sunk on almost every stream and at almost every ferry..."

Arson is the oldest force multiplier, and in communities where wealth consists of livestock and farms, it has always been a traditional weapon of insurgency.

A little historical aside here: when one Fenian conspirator attended the funeral of O'Donovan Rossa, his longtime comrade, the surviving Fenian grumbled to a friend that he was attending under protest, because Rossa "...didn't have the guts to burn a single British haystack." Which implies that "one haystack" is the minimum unit required to prove one's courage, haystacks being lonely and (usually) undefended.

But you don't have to look to other countries for proof of arson's appeal to the disenfranchised, or those who consider themselves so. William Faulkner's great short story, "Barn Burning," is about the career of an ex-Confederate arsonist, patriarch of the Snopes family, who expresses his dislike for his enemies (the whole world, in his case) by burning their barns and other outbuildings.

It's one of the scariest stories ever written. The viewpoint is that of the arsonist's son, who tries to be loyal but constantly hopes that "Father" will settle down and not instantly make enemies of every new neighbor, after they're forced yet again to move on or be lynched for arson. "Father" is good at his hobby, so there's usually no proof, but the charges are always dropped with a warning:

> "I can't find against you, Snopes, but I can give you advice. Leave this country and don't come back to it."

Faulkner does a brief flashback to "Father's" time in the Confederate Army, and it's definitely not the usual Lost-Cause narrative:

> "His father turned, and he followed the wiry figure, walking a little stiffly from where a Confederate provost man's [military police] musket had taken him in the heel on a stolen horse thirty years ago..."

After loading up the wagon and fleeing, "Father" builds a very small fire for the night. Faulkner makes fire into "Father's" god, a survival of his days as a Confederate deserter:

> "That niggard blaze was the living fruit of nights passed during those four years in the woods hiding from all men, blue or gray, with his strings of horses (captured horses, he called them)...The element of fire

spoke to some deep spring of his father's being...as
the one weapon for the preservation of integrity, else
breath were not worth the breathing..."

The same impulse naturally came to other outsiders, with a better
cause than Snopes's: Fire is a way for a powerless individual strike
against the community and, in theory, survive to strike again.

Confederate POWs who escaped northern prisons and fled to
Canada had the same idea: Arson was the only way that a few infiltrators (with plenty of looking-the-other-way from Canadian authorities) could hope to destroy the gigantic Yankee cities. Some of these
Confederate infiltrators formed the Confederate Army of Manhattan
with the avowed purpose of burning down Manhattan.

There were only eight soldiers in this "army" led by an ex-Geor-
gian named Robert Cobb Kennedy. They had petroleum-based ac-
celerants, lighters, and a plan to start simultaneous fires in Manhattan's biggest hotels. You'd figure they couldn't fail. Manhattan in the
mid-19th century had so many massive fires, without the benefit of
dedicated arsonists, that two different conflagrations ten years apart
got the title "Great Fire of New York." In 1835, just 26 years before the
Civil War, a fire burned 17 blocks of Manhattan's financial district.

Ten years later, in 1845, there was another "Great Fire of New
York." Once again, a big chunk of the Financial District was destroyed.

But fail the Confederate saboteurs did, somehow, and skedaddled
back to Canada. It's not clear why they failed—or rather, "HOW? IN
GOD'S NAME, HOW?" Most cities in Europe, as well as the US,
were not so much cities as bonfires-in-waiting, needing only a match
and a wadded newspaper to become a merry blaze.

But somehow these crack secret saboteurs managed not to ignite anything worse than a few trash fires. It was a sad day in the history of arson campaigns.

Perhaps it was a personnel problem. The group's leader, Kennedy, was described as "unprepossessing," a Victorian euphemism for "the guy looks like a complete fool."

With all the crowded buildings in Manhattan, Kennedy decided to burn Ripley's Museum. Why? Because he thought it would be "fun." This is the kind of recruit who would be described in the Commonwealth as "not out of the top drawer."

Worse yet, Kennedy failed even at this self-appointed task. Believe it or not, he failed to burn down Ripley's Museum. None of the other seven soldiers of the CAM (Confederate Army of Manhattan) managed to burn down their assigned hotels. They should have paroled Faulkner's barn burner and put him in charge, instead of shooting him in the foot.

But, let's face it, to be strategically effective arson has to be carried out on a massive scale, with a real army. And that's what happened in the last year of the war, as the North realized that the Confederates would not yield to reality and surrender. In 1864-65, arson became Union policy, in Atlanta and rural Georgia during Sherman's March to the Sea and up through the Carolinas, and in the Shenandoah Valley in Virginia, under Grant and Sheridan.

During Sherman's march north through South Carolina, the state capital, Columbia, was burned. The burning of Columbia, which probably wasn't the fault of Union forces, is still a deep grudge for some South Carolinians.[2]

"Americans have no sense of history"? Nah, the wrong ones have way too much of it.

Sherman probably didn't burn Columbia, but he should have. The great arsons of the late war years were common-sense military measures, and should have been done years before, and the incessant wailing over the burning of Columbia is especially puerile because Columbia was and is still the capital of the state where the Civil War was fomented.

Everyone in the Confederacy agreed that South Carolina was the most fiercely pro-war, pro-secession state, home of the fire-eaters. By any metric you choose, SC needed to be slapped into sanity, because there was simply no sane faction in the SC polity.

South Carolina was the only state to secede in 1860, long before the others, in fact it had tried to secede as early as 1831. A more interesting stat is that it was the only state of the Confederacy that did not produce a single white Unionist unit in the whole Civil War. The notion that South Carolina had it coming was a common view even in the Confederacy; Georgians whose barns were burned by Union soldiers often said, "Why burn our stuff? Why don't you save it for those SC bastards?"

The other well-known Union act of official arson, the burning of the Shenandoah, was simple military necessity. The Valley had fed the

2. On February 17, 2017, locals dressed in Confederate uniform to protest the "gigantic" warcrime on the 152nd anniversary. "Gathering Held at State House to Remember Burning of Columbia" WLTX.com, February 18, 2017.

Army of Northern Virginia for years of war even while intermittently occupied by the Union. Stripping the fields and barns was the simple solution, and would have been ordered by any army other than the US long before 1864. The British, the Russians, the French, all would have left the Shenandoah in ashes long before Grant did.

Leroy Gresham was dying as those great arson campaigns were carried out. The house fires in Macon during the early war years were forgotten, as whole districts of the Confederacy burned. No one boasted about setting those little fires in Macon after the war, so we still don't know who set them, or even whether there were really more arson fires in those years than in any other year. Once you build a city of dry wood and fill it with open fires, it will eventually burn.

Still, it seems probable that some of the fires that frightened Mary Chesnut and made the invalid Gresham take notice were arsons. They didn't matter much strategically, but they were part of the atmosphere of doom that forced the Confederacy to strengthen the Home Guard and turned Mary Chesnut's social circle into light sleepers for the duration.

Like all guerrilla/insurgent activity, carrying out an arson attack is much scarier and more dangerous than most of us civilians think. The failure of Kennedy's eight stooges in the Confederate Army of Manhattan should demonstrate that. The anonymous arsonists of Macon, GA, during the war years managed, at least, to burn down some of their oppressors' buildings. And this sort of action, a kind of attrition of the will to fight, is one of the guerrilla's first tasks.

Chapter Thirteen

Twain at War

I *f you look at the dates of Mark Twain (Samuel Clemens), you might wonder where he was during the war. The short answer is, he skedaddled.*

Everybody knows Mark Twain, but not everybody knows where he was during The War.

He was born in Missouri in 1835, so he would have been 26 when the Civil War started in 1861. If you were from that generation, and especially if you were a white Southerner, you were going to be asked, in the years after the war, "Where were you during the fighting?" The South's army, which recruited only whites, was outnumbered and so enrolled at least 70 percent of white men "of military age."

That was an awkward question for Twain. The short answer is that he skedaddled for California in 1861. His brother Orion was secretary of the Nevada Territory, which gave Twain a good excuse, but most Americans of that era would have guessed his age, wondered what he did in the war, and on finding out that he had done nothing at all—had spent The War scuffing around California and Nevada—would have concluded that he was a shirker, a coward.

This was a serious disgrace, and a lasting one. For example, James Thurber wrote in the 1930s about his grandfather, a veteran of Sherman's March, and contrasted his grandfather's pride in his service with the family legend about Cousin Zenas:

> "Zenas had died in 1866. A sensitive, rather poetical boy of twenty-one when the Civil War broke out, Zenas had gone to South America—'just,' as he wrote back, 'until it blows over.' Returning after the war had blown over, he caught the same disease that was killing off the chestnut trees in those years, and passed away. It was the only case in history where a tree doctor had to be called in to spray a person, and our family had felt it very keenly; nobody else in the United States caught the blight. Some of us have looked upon Zenas' fate as a kind of poetic justice."

As Thurber's memoir shows, the stigma of dodging The War was a serious matter well into the 20th century. If you were a young man during the war, you were supposed to have enlisted and fought.

Twain hadn't, as far as anyone knew. He became, after the war, a strong Unionist, which would have caused him some trouble among Southerners, but not having fought at all must have caused him many an awkward moment even among northerners in New York and Connecticut, where he lived. Twain finally wrote about his war-dodging history in 1885, when he published "The Private History of a Campaign that Failed."

It's a very good, but very tricky essay, worth reading for its own sake—parts of it are very funny—but also as a good description of

Missouri in 1861, when the war started. Twain made his living as a steamboat pilot on the Mississippi. The river towns were linked to the South by the cotton and slave trades but also crowded with recent immigrants who were fiercely loyal to the Union. It was a confused political landscape, and Twain does his usual excellent job of portraying that confusion.

But he's also justifying his own failure to enlist, making an excuse for surviving, in fact. Not that Twain was likely to have felt what's now called "Survivor's Guilt." He seems to have felt that about something else, his brother's death in a steamboat explosion, but not for dodging the war.

He had dodged the war for what he considered very good reasons, but reasons that weren't easy to explain to contemporaries who'd risked everything to fight.

So the first thing Twain does is to state, like a good orator, why he has a right to speak on the matter. The essay begins:

> "You have heard from a great many people who did something in the war; is it not fair and right that you listen a little moment to one who started out to do something in it, but didn't? Thousands entered the war, got a taste of it, and then stepped out again, permanently. These, by their numbers, are respectable, and are therefore entitled to a sort of voice—not a loud one, but a modest one; not a boastful one, but an apologetic one. They ought not to be allowed much space among better people—people who did something—I grant that; but they ought at least to be allowed to state why they didn't do anything, and to explain the process by which they didn't do anything."

It's a remarkable first paragraph. On the surface it's all modesty, but the modesty is transparently false, because in 1885, when the essay was published, Mark Twain was the most popular literary celebrity in the US—his masterpiece, *Huckleberry Finn*, had just been published—so he had no need to apologize to anyone.

But under this modest tone is Twain's real anxiety about The War. He's in a difficult rhetorical situation, facing hatred from other white Southerners for deserting the Lost Cause, and scorn from Northerners for simple cowardice. The rest of the essay is Twain's attempt to wedge himself a way out of these anticipated reactions.

Buried in this show of modesty is an interesting claim: That quite a lot of young men of military age ("These, by their numbers, are respectable...") dodged the war. Thurber's family shame at Cousin Zenas's cowardice reflects the same memory of young men dodging the war.

Most northern men who didn't want to join the war avoided enlisting by legal means, but quite a lot just made themselves scarce, as Zenas did:

> "Over 160,000 individuals...escaped the call to arms by refusing to report to their draft boards for examination. These men were illegal draft evaders by choice and deserters by law."[1]

---

1. Levine, Peter. "Draft Evasion in the North During the Civil War, 1863-65". The Journal of American History. (P. 816-834) Vol. 67 No. 4. March, 1981.

So Twain is telling an awkward truth when he claims that he had a lot of company, northern as well as southern.

Rather than push this awkward truth, Twain returns to his stylized humble and humorous tone, asking only that he and the others who avoided serving "should be allowed to explain why they didn't do anything...there was a good deal of confusion in men's minds during the first months of the great trouble—a good deal of unsettledness, of leaning first this way, then that, then the other way. It was hard for us to get our bearings."

This is another awkward fact. States like Missouri had a stake in the Union, since they depended, especially in river towns like Twain's Hannibal, on trade up and down the river, which a war would disrupt. Missouri convened a special legislative session and voted 98-1 against secession.

But Hannibal was in Marion County, which was part of a strip of northern Missouri called "Little Dixie," settled by southern pro-secessionists.

Missouri was a violent, confusing landscape during the 1850s. There was a strong pro-Confederate elite, and only by mobilizing the St. Louis German population, who were uniformly pro-Union, could Nathaniel Lyon stop the state's elite from pushing Missouri into the Confederacy.

Twain illustrates the confusion of 1861 with a story about his experiences with a fellow cub pilot. The story may be true, and it may not, but Twain generally knew pretty well what he could get away with, so it certainly rang true for his audience:

> "My pilot-mate was a New Yorker. He was strong for
> the Union; so was I. But he would not listen to me
> with any patience; my loyalty was smirched, to his

eye, because my father had owned slaves. I said, in palliation of this dark fact, that I had heard my father say, some years before he died, that slavery was a great wrong, and that he would free the solitary negro he then owned if he could think it right to give away the property of the family when he was so straitened in means. My mate retorted that a mere impulse was nothing—anybody could pretend to a good impulse; and went on decrying my Unionism and libeling my ancestry."

The New Yorker soon changes his mind, in lockstep with his community:

"A month later the secession atmosphere had considerably thickened on the Lower Mississippi, and I became a rebel; so did he. We were together in New Orleans, the 26th of January, when Louisiana went out of the Union. He did his full share of the rebel shouting, but was bitterly opposed to letting me do mine. He said that I came of bad stock—of a father who had been willing to set slaves free."

Twain's point, quietly insinuated, is that he and his co-pilot had no minds of their own worth mentioning; that they changed their minds with the prevailing "atmosphere." That's a cynical but accurate description of how most men of military age made their choices.

Writing in 1885, Twain can count on a broad, if lukewarm, consensus that slavery was "a great wrong," so his friend's willingness to smear his ancestry because his "father had been willing to set slaves free" makes the friend's changing stance increasingly ludicrous. So this entire long illustration is not simply showing his friend's fatuousness, but implying that there was no moral sense undergirding such decisions.

Twain finishes the anecdote with the third and final stage of his friend's political evolution:

> "In the following summer he was piloting a Federal gun-boat and shouting for the Union again, and I was in the Confederate army. I held his note for some borrowed money. He was one of the most upright men I ever knew; but he repudiated that note without hesitation, because I was a rebel, and the son of a man who owned slaves."

Twain implies several grim lessons via this anecdote. First, that his friend's allegiances were a matter of conforming to the prevailing "atmosphere," so that when the Union won the battle for control of the Mississippi, the friend turned Union and, not coincidentally, got a job as a pilot again. That was a big deal, because as Twain explains at length in Life on the Mississippi, piloting was by far the best-paid job you could get in his part of the world.

Second, Twain begins to imply a distinction between his own hapless allegiances and those of his friend—because he chose the Confederate Army, which was never very successful on the Mississippi or its riverbank towns.

So Twain, though equally stupid, seems a little less venal than his friend.

Finally, he says his ex-friend welched on money he owed Twain. Sure, Twain says very clearly that his friend was "one of the most upright men I ever knew," but Marc Antony said some nice things about Brutus in much the same way, with the same intention. You aren't meant to take Twain's defense of his friend seriously; you're meant to think that his friend, who's already been revealed as a facile conformist, couldn't have been all that "upright" and, like those who say, "it's not the money, it's the principle," was really just cheap and dishonorable.

Of course, this sort of fickle allegiance-switching didn't happen everywhere. If you were from Maine, you were going to fight for the Union if you fought at all. If you were from South Carolina, you were going to fight for the Confederacy.

If you were from Missouri, Kentucky, Maryland, or Tennessee, you might choose either side, but even in those states, the choice was rarely a matter of personal opinion. More often, it was a regional decision. East Tennessee meant pro-Union, "Little Dixie" towns in Missouri (like Hannibal) meant Confederate, while St-Louis German meant Union. But it's worth remembering that every state of the Confederacy, except South Carolina, had at least one unit of white residents serving in the Union army.

Units of the USCT (African-American ex-slaves) were raised in SC late in the war, but not one unit of white South Carolinians was formed in the whole course of the war. That state was something special, a very strange place.

Twain, as a child from the Little Dixie part of Missouri, would naturally follow the local herd of cattle into the Confederate ranks, as he confesses he did. He goes on to confess, in this essay, that he tried to be a Confederate for only a month, then lit out for the West. This puts him in an awkward spot in 1885, confessing to having tried to fight for the losing side, and worse having deserted from it as well.

The rest of the essay relies on the points made in the story of Twain's changeable friend: that he and the rest of his demographic were young, stupid, fickle, and generally deluded. The best thing about this characterization of his demographic is that stupid, fickle, deluded people are great comedy material. Twain relies on this comedy to keep his audience, likely to be wary of a draft-dodger's self-justifications, sufficiently entertained to go along with him.

Is Twain's account of his service true? I don't know, and AFAIK no one else does. Most commentators simply accept it as true, a dangerous move.

My guess is that they're like the stories in *Papillon* or other popular memoirs: some literally, autobiographically true, some borrowed from other people's stories, some just plain made up.

Most likely, the bulk of it is more or less true, just because Twain was a born storyteller and such people generally start with the more slapstick events in their past and riff them up a bit.

For example, Twain tells a brilliant story about the name his comrades chose when they formed their Confederate volunteer unit. This story will seem familiar to those who remember Huck's account of the "Pirate Band" in *Huckleberry Finn* (Chapter 2, "Our Gang's Dark Oath") but that doesn't mean it's necessarily made-up. He just had a gift for that kind of take-down:

> "...we called ourselves the Marion Rangers. I do not
> remember that any one found fault with the name. I
> did not; I thought it sounded quite well. The young
> fellow who proposed this title was perhaps a fair sam-
> ple of the kind of stuff we were made of. He was

young, ignorant, good-natured, well-meaning, trivial, full of romance, and given to reading chivalric novels and singing forlorn love-ditties. He had some pathetic little nickel-plated aristocratic instincts, and detested his name, which was Dunlap; detested it, partly because it was nearly as common in that region as Smith, but mainly because it had a plebeian sound to his ear.

"So he tried to ennoble it by writing it in this way: d'Unlap. That contented his eye, but left his ear unsatisfied, for people gave the new name the same old pronunciation—emphasis on the front end of it. He then did the bravest thing that can be imagined,—a thing to make one shiver when one remembers how the world is given to resenting shams and affectations; he began to write his name so: d'Un Lap. And he waited patiently through the long storm of mud that was flung at this work of art, and he had his reward at last; for he lived to see that name accepted, and the emphasis put where he wanted it, by people who had known him all his life, and to whom the tribe of Dunlaps had been as familiar as the rain and the sunshine for forty years."

Ah, what fine insidious comedy in this story! It gets the laughs while reinforcing Twain's very grim thesis about his demographic: "young, ignorant, good-natured, well-meaning, trivial, full of romance, and given to reading chivalric novels..."

Above all, Twain implies that people are so stupid that they can disable their memories, as when people "...to whom the tribe of Dunlaps had been as familiar as the rain and the sunshine" when faced with a Big Lie—that the name is actually an ancient, aristocratic French one, "d'Un Lap."

Twain's hatred for "chivalry" as it was understood in the South shows up in everything he wrote, from *Huckleberry Finn* to *A Connecticut Yankee in King Arthur's Court*. He names the wrecked riverboat in *Huck Finn* the Sir Walter Scott (Chapter 8, "Honest Loot from the Walter Scott") , and his hate for Scott's "chivalrous" novels was white-hot:

> "Then comes Sir Walter Scott with his enchantments, and by his single might checks this wave of progress, and even turns it back; sets the world in love with dreams and phantoms; with decayed and swinish forms of religion [Not fond of Cat'lics, Twain wasn't]; with decayed and degraded systems of government; with the sillinesses and emptinesses, sham grandeurs, sham gauds, and sham chivalries of a brainless and worthless long-vanished society. He did measureless harm; more real and lasting harm, perhaps, than any other individual that ever wrote." *(Life on the Mississippi, Chapter 46)*

That's why Twain ends *Connecticut Yankee* by electrocuting the entire "chivalry" of England in their iron armor. It's meant to be a happy ending.

And of course it's not fair at all to Scott, a hard-working and competent novelist. And if one were to make a serious attempt to rank Victorian novelists by the harm they did... well, where would you even start?

It's a tough call. The Empire enforced its domination on Ireland, India, Africa, Asia. Do Thackeray's vicious caricatures of these oppressed people count for more than Dickens's strategic silence about imperial abuses? Hard to say, but silence is generally a better cover for mass death than vituperation.

Compared to either of those writers, Scott seems relatively harmless—as long as we're talking about the effect of his work in Britain. But Twain isn't really concerned with Britain. What he hates so violently is the effect of Scott's novels in the American south.

Scott's novels had their worst effects in Dixie. They helped to convince a small, powerful clique of lowland farmers who got obscenely rich off slave labor that they were "cavaliers," a term they borrowed from Scott's novels without a clue about its real contexts.

It's a shame that these South Carolina cotton farmers couldn't have socialized with the real 17th-century Cavaliers. The snubs would have been violent enough to raise a mushroom cloud over Charleston. One of the implications here is that delusion of genuinely influential Charleston plutocrats was far more absurd and destructive than that of Twain and his friends imagining themselves as Marion rangers.

As it was, the planter class of the South invented itself as the uninvited heirs of the Cavaliers, and extrapolated from it that they were braver, nobler, finer in their manners and above all better fighters than the Puritans of the North. Which was odd, because the actual Roundheads crushed the actual Cavaliers time after time—but you often get these reactionary cliques idolizing losing armies, like the Wehrmacht and, a few decades after its surrender, the CSA itself.

Twain's "Marion Rangers," march out of Hannibal, looking for the war. They never quite find it, partly because they don't really want to.

Twain calls them a "herd of cattle," and he means it, because his grim thesis throughout the essay is that people are too stupid to be held to account—a herd species, gregarious to a fault, like cattle:

> "What could you expect of them? They did as well as they knew how, but really what was justly to be expected of them? Nothing, I should say. That is what they did."

Twain is not kidding at all. He always spoke of his work as "preaching," and what he preached was something close to nihilism. The humor was the spoonful of sugar that helped this nasty-tasting medicine go down, as he said himself:

> "I have always preached...If the humor came of its own accord and uninvited, I have allowed it a place in my sermon, but I was not writing the sermon for the sake of humor. I should have written the sermon just the same whether any humor applied for admission or not."

Twain's sermon in this essay sweetens up its audience for a while, then gets to the point when the Marion Rangers kill a stranger coming down the road at night, out of fear. They don't know who he was, and they weep over the body, as stupid and sentimental as ever.

Twain does a beautiful job of changing tone here. One minute he's telling the story of his comrades' battle with the rats in the corn-crib where they were staying:

> "...the place was full of rats, and they would scramble over the boys' bodies and faces, annoying everybody, and now and then they would bite someone's toe, and the person who owned the toe would...throw corn in the dark. The ears were half as heavy as bricks, and when they struck they hurt. The persons struck would respond, and inside of five minutes every man would be locked in a death-grip with his neighbor. There was a grievous deal of blood shed in the corn-crib, but this was all that was spilt while I was in the war."

You can imagine the audience's delight with this typical Twain story, just a bunch of lunkheads having a corn-cob fight in the dark. That's where he wants his audience when he zaps them with this transition:

> "No, that is not quite true. But for one circumstance it would have been all. I will come to that now."

There's nothing as disarming as a good laugh, and Twain wants the audience disarmed at this point, because everything is going to change. He's riffed hilariously on the Marion Rangers' cowardice, ineptness—it was news to me that small-town youths like these didn't know how to ride horses, for example:

> "We were town boys, and ignorant of horseman-
> ship...We did learn to ride, after some days' practice,
> but never well."

But the humor is always at the service of the sermon, and the
sermon starts with that deft transition about the one other time that
blood was shed, the non-humorous incident.

The Hannibal boys have ducked possible combat many times,
moving from place to place every time they hear a rumor that Union
troops were in the area, prompting one farmer to remark, "Marion
Rangers! Good name, b'gosh!" and suggesting that they "...could be
counted on to end the war in time, because no government could
stand the expense of shoe-leather it would cost trying to follow us
around."

It's this sheer cowardice that leads them to murder a stranger.
They're always hearing rumors of Union advances, but they eventually
stop taking them seriously, just because they're worn out with fleeing,
and decide to stay where they are: "...the enemy was [said to be] hov-
ering in our neighborhood. We said let him hover."

Then they get scared again, and take up watch in the darkness. They
can hear hoofbeats, and see:

> "...a man on horseback; and it seemed to me that
> there were others behind him...Somebody said 'Fire!'
> I pulled the trigger. I seemed to see a hundred flash-
> es and hear a hundred reports, then I saw the man
> fall down out of the saddle. My first feeling was of
> surprised gratification; my first impulse was an ap-

prentice-sportsman's impulse to run and pick up his game."

The Rangers wait for the attack of the other riders, and soon realize there were none. They've killed a lone horseman. They creep up to the body:

> "When we got to him the moon revealed him distinct-
> ly. He was lying on his back, with his arms abroad;
> his mouth was open and his chest heaving with long
> gasps, and his white shirtfront was all splashed with
> blood."

This is the moment everything in the essay shifts radically. As Melville wrote after the battle of Shiloh, "What like a bullet can undeceive?"[2]

The Marion Rangers are undeceived by their own volley of shotgun pellets, and switch from fierce soldiers to mourners in a second:

> "The thought shot through me that I was a murderer;
> that I had killed a man—a man who had never done
> me any harm...I was down by him in a moment, help-
> lessly stroking his forehead; and I would have given
> anything—my own life freely—to make him again
> what he had been five minutes before. And all the boys
> seemed to be feeling in the same way; they hung over

---

2. "Shiloh, A Requiem" (April, 1862) by Herman Melville.

him, full of pitying interest, and tried all they could to
help him, and said all sorts of regretful things."

The Rangers never find out who the man was—or at least that's the
way Twain tells it. No one in the area knows him. He wasn't wearing
a uniform, and wasn't armed.

Now Twain is ready to preach his sermon without disguise:

> "The thought of him got to preying upon me every
> night; I could not get rid of it...And it seemed an
> epitome of war; that all war must be just that—the
> killing of strangers against whom you feel no personal
> animosity...My campaign was spoiled."

That short, blunt sentence, "My campaign was spoiled," recalls
the title of the essay and echoes back, through the seemingly modest
opening paragraphs, forcing the reader to reevaluate Twain's humble confession that he "did nothing." If war is murdering random
strangers, then doing nothing is the better choice.

But the sermon detonates in the reader's mind only after Twain's
long and very funny account of his slapstick early service. There's
plenty of room for the audience, which may include maimed veterans,
prone to outrage, the respect they're accustomed to get: "It seemed to
me...that war was intended for men, and I for a child's nurse."

Twain ends the essay with another punchline—one of the weakest
in the essay:

> "There was more Bull Run material [i.e. rookie sol-
> diers prone to flight] than exhibited itself at Bull Run.

> And yet it learned its trade presently, and helped to
> fight the great battles later. I could have become a
> soldier myself, if I had waited. I had got part of it
> learned; I knew more about retreating than the man
> that invented retreating."

That's the end of the essay, and the feebleness of the joke is intriguing. It ranks last, a distant last, in all the laugh lines of the essay. Some of the others are truly hilarious, like the story of the Rangers' battle with the farmer's dogs. Yet Twain ends with this formulaic line about retreating.

When a writer as good as Twain saves, not his best, but his worst punchline for last, you have to wonder why.

My own guess, and it's just a guess, is that Twain the preacher, who happened to be a gifted humorist, wants to send the audience away with a joke, but not a very good or memorable joke, because as he confessed, the sermon itself is his real point. He's a humorist; that's his job, and he has to end with a joke. But having turned the battle from a humble self-defense for his own cowardice into the strong implication that shunning the war was in fact the nobler course, he doesn't want that final joke to overwhelm the sermon. It's a marker to the audience, "Here, here's your obligatory final punchline; here endeth the sermon."

So the essay ends with an intentionally vague message. If you were in a Northern, pro-Union audience (as Twain's primary audience was when he wrote this essay), would you feel that Twain was criticizing you and your relatives as well as his own Hannibal "town boys"? Would an audience that cherished the memory of Gettysburg feel that they shared the blood guilt of the Marion Rangers?

I suspect not. There are too many pointers in the essay suggesting that blame for the war veered south. First, the use of "slavery" as the test issue in his long opening anecdote about his fickle fellow pilot. There's no sign that this colleague had any sense of the immorality of slavery, but Twain's family clearly did. He never varies on that point.

Then there's the characterization of the dimly seen Union forces. Twain mentions that one such force was led by Ulysses Grant, whose memoirs Twain had edited (and co-written, if you ask me).

Finally, there's the effect of the many tales of Walter Scott-like fancy, exemplified by Dunlap, aka "d'un Lap," "the mongrel child of philology," who invents the name "Marion Rangers" and continues his fanciful naming of every camp the Rangers occupy in their endless retreat: "[Dunlap] proved useful to us in his way; he named our camps for us, and he generally struck a name that was 'no slouch,' as the boys said."

This impulse to impose a chivalrous, ridiculous nomenclature on the landscape of rural Missouri was the same one that Twain excoriated, to the point of abandoning all attempt at humor, in his loathing for Sir Walter Scott's role in the mass delusion that Twain considered the Confederacy to have been.

This idea was clearly in his mind in the 1880s. It's the theme of one of the grimmest and least amusing chapters of Huckleberry Finn, Chapter 12, in which Colonel Sherburn, after killing a harmless drunk who offended his southern honor, faces down a would-be lynch mob and addresses them at length, denouncing their delusion that they're a brave, fierce people and dismissing them with a contemptuous "Now git!"

So Twain probably meant the essay to resonate in a carefully circumscribed manner, sparing the Unionists, who had never boasted as the proto-Confederates had and had gone to war in a less deluded, less euphoric mood, as something they considered duty but were not especially looking forward to.

But the essay clearly means to echo beyond the Confederacy. Twain's Rangers are, as he says, "a fair sample" of the human material with which wars are made: "There were scores of little camps scattered over Missouri where the same thing was happening."

And he makes clear, several times in the essay, that, ridiculous as they were in their rookie outing, they learned the trade of war all too well:

> "One might justly imagine that we were hopeless material for war. And so we seemed, in our ignorant state; but there were those among us who afterward learned the grim trade; learned to obey like machines; became valuable soldiers; fought all through the war, and came out at the end with excellent records."

In fact, not all of the Rangers "came out at the end" of the war at all. One of the Hannibal boys, introduced as a "vast donkey," turned into such a good soldier that "he stuck to the war, and was killed in battle at last."

It's a remarkably deft essay, navigating a course trickier than any low-water river channel Twain ever piloted in his youth, and doing it with some of the funniest war stories ever told. Still, you can see why Twain ended up cursing life in general and recommending that souls in danger of being born choose oblivion while they can.

Chapter Fourteen

Chernow's Grant: The Big Pithed Plinth

There should be room for a new, merciless, biography of Grant. Grant is a strange figure: noble yet resentful and petty. There are so many legends about him that don't seem to add up, like his drunkenness, his failure. You'd think a new biography would take those topics on, but alas, we get Ron Chernow handing in another civics text.

"They think that things are all right/ Since the deer and the dachshund are one." *W.S. Stevens "Loneliness in Jersey City"*

Grant's memoirs are superb and shouldn't require any commentaries, but the commentaries continue to appear because there's a strong market for new biographies ripping Grant out of his time and into the ephemeral fancies of our own.

A lot of Grant's current wave of fame is due to a 400,000-word biography of him by Ron Chernow published in 2017. I haven't been able to finish it, which is rare for me. I'll read just about anything on my beloved Civil War, unless it's pro-Confederate bathos—but I shied from Chernow's Grant bio without quite knowing why. It's something you develop if you've wasted your life hiding among the bookshelves, a "map of fear" warning you away from some text-forests. Sometimes your fears prove wrong, and you've slighted a worthy book, or even a whole genre.

Sometimes. Not this time.

Instead, when I buckled down to reading Chernow's book, what hit me was what old-fashioned American piety it exudes. There's a whiff of public-school classrooms in Chernow's story, the old smell of chalk dust and baloney sandwiches, the perfume of a sixth-grade Social Studies class.

This is the unreconstructed version of U.S. Grant—the same Grant I once learned to worship from mid-20th-century boy's books by Bruce Catton, de facto laureate of the Civil War Centenary years. Far from making a new Grant, Chernow has delivered his 400,000-word load of prose concrete to firm up the humble, simple Grant who inhabited the mainstream pantheon in the Eisenhower years.

Here are a couple of typical Chernow passages describing Grant's parents:

> "After hanging out the shingle for his own tannery in Ravenna, Ohio, Jesse [Grant, father of U.S. Grant] contracted a 'debilitating fever and ague' that forced him to discontinue business...

"Full of vim and drive, Jesse searched for a worthy
partner and found her in the pious, frugal Hannah
Simpson..."

There's a word for this kind of prose. No, not that word. I was
thinking of "quaint." So quaint as to verge on "camp." It's a collection
of overpriced antique vocabulary items, from "hanging out the shingle" to "vim and drive" and "pious, frugal Hannah." Imagine someone
you know using "vim" in conversation, or describing a Evangelical
mom as "pious" and "frugal."

These might be the words Grant's contemporaries would have
used, though only thoroughly urban people imagine that provincials
would be so generous in describing their neighbors—but it's a strange
form of English to issue from the keyboard of Ron Chernow.

A quick look at Chernow's bio illustrates the gap between the author's true dialect and this Pepperidge Farms quaint style. Chernow's
father created a stock brokerage firm in NYC; young Chernow was
voted "Most Likely to Succeed" at his elite high school and graduated
summa cum laude from Yale and went on to Cambridge.

Chernow is a classic post-war American blue-chipper. The problem with these people is that, as Wallace Stevens said, "They think
that things are all right." This background doesn't make Chernow
incapable of portraying Victorian provincials like Grant, the point
here is that he really doesn't want to know that they were different
from his idea of a good American. This makes his attempt to mimic
their diction and attitudes mawkish and patronizing.

Chernow would not stick with a doctor who diagnosed "ague"
any more than he would with a lawyer who "hung his shingle" any-

where outside midtown Manhattan. It's insulting to be patronized by language like this by an author writing a full century after Lytton Strachey showed how to portray the demonic yet awe-inspiring reality of Victorian celebrities like Grant in his biographical essay, *Eminent Victorians*.

Eminent Victorians is a brilliant study of four Victorian celebrities: Cardinal Manning, Florence Nightingale, Charles Gordon, and Thomas Arnold, which reveals them as warped, driven, ambitious products of a culture obsessed with the striving for fame. Strachey is not shy about using words like "demon" to characterize the Victorians, even in his sketch of the revered Florence Nightingale:

> "The Miss Nightingale of fact was not as facile fancy painted her. She worked in another fashion, and to-wards another end; she moved under the stress of an impetus which finds no place in the modern imagination. A Demon possessed her. Now demons, whatever else they may be, are full of interest. And so it happens that in the real Miss Nightingale there was more that was interesting than in the legendary one; there was also less that was agreeable."

Grant had several demons clawing at his insides, driving him to drink, or more often, to envy and bitterness. But Chernow's not interested in dissecting the tricky stuff. His hokey prose affectations might be a minor annoyance, if they didn't match his determination not to look too hard at the murkier aspects of Grant's career. There's nothing new here—and yet the success of Chernow's earlier Hamilton bio as well as the Grant book shows that he's fulfilling some need for his American elite audience.

It's not that Chernow needs to debunk Grant. Grant was a brilliant commander, the closest thing the North had to a general who understood Gurowski's policy of just going for the Confederacy's throat. Left to himself, in fact, Grant would have ended the war much more quickly, with far fewer Union casualties, had not the Washington elite vetoed his plan to take Richmond using naval power coming in from the Atlantic side.

But Grant's ability to play the elite, to see how political power was tilting and exploit it for his (and his army's) benefit, was a big part of his success. Grant was not naïve, to put it mildly, and he doesn't need to be resculpted in 20th-century Rushmore style. Especially not at such absurd length.

The rule for biographers should go something like, "If you don't have anything new to say, don't take almost a half million words to re-say the old version."

My first red flag was the fact that Chernow accepts Grant's judgment of his colleagues. This is fatally naïve. Grant was a fine man, a great general, but also a ruthless courtier. This is made very clear in his cruel libels against General George Thomas, a figure worshipped by true Civil War buffs.

For those not familiar with the old canard about Thomas, it goes like this: George Thomas was a good defensive general but a real slow-poke on offense. He was like an Offensive Tackle who can pass-block at All-Pro level but isn't much use at opening holes for the runner.

And where did that depiction come from? Above all, from the pen of U.S. Grant, who damned Thomas with very faint praise in his *Memoirs*. Yes, Grant, the simple, good man of legend, was a good hater. You wouldn't know it from Chernow's book, but Grant could hold a grudge with the best of 'em. And Thomas was one of his biggest targets.

Being a brilliant and twisted Victorian, Grant was well aware he'd been a little hard on Thomas, who may well have been the best Union general of the war. Here's a passage from Grant's *Memoirs*, showing not just his hostility to Thomas but his rhetorical sophistication in modulating it:

> "As my official letters on file in the War Department, as well as my remarks in this book, reflect upon General Thomas by dwelling somewhat upon his tardiness, it is due to myself, as well as to him, that I give my estimate of him as a soldier.... I had been at West Point with Thomas one year, and had known him later in the old army. He was a man of commanding appearance, slow and deliberate in speech and action; sensible, honest and brave. He possessed valuable soldierly qualities in an eminent degree. He gained the confidence of all who served under him, and almost their love. This implies a very valuable quality. It is a quality which calls out the most efficient services of the troops serving under the commander possessing it.

> "Thomas's dispositions were deliberately made, and always good. He could not be driven from a point he was given to hold. He was not as good, however, in pursuit as he was in action. I do not believe that

he could ever have conducted Sherman's army from
Chattanooga to Atlanta against the defenses and the
commander guarding that line in 1864. On the other
hand, if it had been given him to hold the line which
Johnston tried to hold, neither that general nor Sherman, nor any other officer could have done it better...
Thomas was a valuable officer, who richly deserved, as
he has received, the plaudits of his countrymen for the
part he played in the great tragedy of 1861–5."

This is not the prose of a simple provincial, or a naïve good guy.
Chernow's Grant comes off almost as an idiot savant—Forrest Gump
or Rain Man in a Union overcoat, while the real Grant was at the other
end of the spectrum: as manic as Sherman, as depressive as Lincoln, as
driven and self-loathing as every other super- and sub-human Victorian hero of the type Strachey dissected.

Chernow doesn't even seem to have absorbed Grant's modulation
when it comes to caricaturing Thomas. He actually throws in the
old joke about "slow-trot" Thomas without explanation, by way of
contrasting Thomas's sloth to Grant's, er, "vim":

"Grant's critique of another older cadet, George H.
Thomas, later hero of Chickamauga, prefigured his
impatience with Thomas's lumbering style of command. [quoting Grant's *Memoirs*] 'At West Point,
when he was commanding cadets at drill, he would
never go beyond a slow trot...The boys used to call
him 'Slow Trot' Thomas.'"

That's a bizarre paragraph. First, the casually tossed-out adjective "lumbering" as a description of "the hero of Chickamauga" makes no sense in any terms. If Thomas had been a "lumbering" commander, he could never have transformed the broken Union fragments at Chicka-mauga into an effective defensive line in mid-battle-time, which is something like real-time cubed. You can't use "hero of Chickamauga" and "lumbering" to describe the same person if you have any claim to do serious military history.

But beyond that, the way that Chernow takes Grant's account of Thomas's nickname at West Point as Gospel is sloppy historiography. Thomas was called "slow trot" by his soldiers because he spent the entire war in physical agony. Just before the war, he stepped off a train to smoke a cigar and fell down a steep gully, damaging his spine for life. He couldn't stand, walk or ride without constant pain. He never mentioned it, never complained, but he simply could not move at anything more than a painful walk, or a slow trot on horseback. Grant knew that, but there's no evidence that Chernow bothered to learn it.

Chernow takes Grant's anecdote about Thomas's alleged slowness at West Point, decades before his spinal injury, as gospel. This is despite the fact that Grant's dislike for Thomas was known by everyone in the Union forces. I don't know whether Grant was slandering Thomas consciously or not but at the very least Chernow could have mentioned Thomas's spinal injury. Try sleeping on camp beds through four years of war over the vast distances of the Western theatre and see how you enjoy anecdotes about your slow trot or lumbering style.

That's the thing about Chernow: he's as trusting as a D.A. transcribing prison gossip to get a conviction, which means some people get taken down as a kind of collateral biographical damage.

Grant's *Memoirs* are one of the finest biographies in American literature. I once taught a whole course with a reading list of three books:

Grant's *Memoirs*, Emily Dickinson's *Collected Poems*, and Whitman's *Leaves of Grass*. But the *Memoirs* are not to be taken as a simple deathbed confession. Grant was tenacious, sometimes to a fault, as the second wave at Cold Harbor learned. And he was determined not to die before finishing his apologia, partly to provide some money for his family, since he'd been swindled out of most of his assets. The swindling resulted from Grant's lifelong partiality, which meant he spent most of his life dividing the world into friends and enemies, with endless trust for friends, no forgiveness for enemies, and very little grey area.

The man who swindled Grant had been classed as "friend"; Thomas was an enemy. And that meant that every mention of Thomas in Grant's *Memoirs* is narrated so as to belittle Thomas. And Grant was very good at it, very sophisticated. Compared to him, Lytton Strachey was a maniac with a machete when it came to cutting Victorians down to size.

Grant might even have had the help of his "publisher," Mark Twain, in writing these *Memoirs*. There certainly are a lot of echoes of Clemens's style in the book, and since Twain and Grant worked together on it every day while Grant was dying of cancer, it's reasonable to suppose that Twain may have touched up the narrative here and there.

The result is, I repeat, a great book, one I used to be able to quote from memory (before old age started picking off the little grey cells as if they were Hood's men at Franklin). But it is not a book to take as gospel in assessing rivals like Thomas.

It's not hard to find information about Thomas's spinal injury. Most Civil War fans know all about it, and it's been decades since serious historians of the war have written Thomas off as casually and unfairly as Chernow does. The truth is, it's just lazy research.

Laziness is an odd charge to make against someone like Chernow, whose Wiki frankly annoyed the life out of me because he seemed such a perfect example of the lifelong over-promoted wonk whose one real talent is "work habits." He's the enthusiast, the consensus-endorsing blue-chipper that everyone knew was "most likely to succeed" even at his top-level high school. His book is 400,000 words long! How can you accuse someone like that of laziness?

Perhaps "self-serving credulity" would be a better description of Chernow's failing. Perhaps that would be a good term for the failing of the whole wonk elite that enjoys his books so much. I'm sure the wonks who planned the Iraq invasion were anything but "lazy" in their habits. I'm sure they worked 18-hour days polishing every page of their plans. But they were lazy beyond belief in their trust in unreliable narrators, their willful ignorance of inconvenient history, and their faith in volume: "binders" of detail in support of a plan that would have been dismissed as ridiculous if summarized in one paragraph of clear prose.

Is that unfair? Compared to what, calling George Thomas "lumbering"?

Admittedly, Chernow's is a strange kind of laziness. If you want the traditional kind of sloth, you're looking at it (raises hand). But my kind of laziness never worked very well in the American system, which rewards enthusiasm, collegiality, and this odd credulity about its most vulnerable and cherished myths. It was Chernow's cheery biography of Alexander Hamilton, after all, which inspired the Broadway hit *Hamilton* (2015), which made the grim statist into a hero in the upper-middle class East Coast demographic. That milieu not only accepts but desperately yearns for biographies which prove to their satisfaction that every American figure, of whatever era, feels just like they do at heart. And realizing this, I began to get a sense of why this ACELA-Corridor elite might have become so infatuated with Chernow's biographies.[1]

Chernow breathes the air those people breathe, and he gave them the Hamilton they wanted. That's certainly what he does with Grant in this ponderous tome. (Yes, "ponderous tome"! There are times when such terms, like tactical nukes, may be legitimately deployed.)

You might think that Chernow would offer this allegedly liberal wonk-dom a wholly renovated Grant, one suited for the DNC's new century. That would be a Grant who's on-board with the progressive line on a host of social changes.

Not at all. Not in the least. That's the big shock of the book: how fiercely it strives to recreate a lost consensus narrative that, like Norman Rockwell's painting of the Little Rock anti-integration riots, "The Problem We All Live With", uses a familiar, obsolescent, sentimental, credulous bag of tricks to add one or two small updates and modifications to a very conservative portrait of America.

Chernow's updates focus on Grant's notorious order expelling Jews from the front, and give slightly more attention to his anti-slavery, pro-African American views. But these adjustments are very minor. The Grant you meet here is Bruce Catton's Grant to a degree which is just bizarre.

The only time Chernow shows real defensiveness about his hero is on the issue of Grant's alcoholism. This seems like a weird thing to worry about it. What the hell does it matter if he was an alcoholic? So

---

1. For a more critical, contextualized portrait of Hamilton, see William Hogeland's *The Hamilton Scheme: An Epic Tale of Money and Power in the American Founding.* (Farrar, Straus & Giroux, 2024).

were two-thirds of the officer corps on both sides, and most civilian men who could afford it. But this worries Chernow; that, and Grant's ethnic slurs—but nothing else, especially not Grant's malice toward rivals, let alone the question of Grant's decisions at Shiloh or on the Virginia front.

Every time Grant's drinking comes up, Chernow balances it with something about the soppy-sentimental bond between Grant and his wife Julia, or Grant's affection for his sons. It's like watching a *Home Improvement* rerun: Dad may say dumb stuff or drink too much beer but he's a good family man. As always, Chernow is willing to make his subject seem like a dullard, as long as he embraces family values, as seen by the American upper-middle class of the early 21st century.

Chernow's odd partiality comes through in the most trivial domestic incidents, which loom very large for him and his readers. At one point, Grant, posted far from his fiancée Julia, floods her with letters and gets only eleven replies over a period of 20 months. Chernow muses:

> "One wonders whether, to test Grant's fidelity or undermine their relationship, [her father] forbade Julia from writing more often. Another possibility is that her chronic eye problem converted even simple letter writing into an onerous task."

These are far-fetched, try-hard conjectures. Chernow grudgingly admits a third, less-virtuous possibility, that Julia's father "…kept Julia busily distracted with St. Louis parties where she would be exposed to hordes of handsome young bachelors"—and in the process attributes Julia's posited fickleness solely to her evil dad.

It's hard to say which of these hypotheses is silliest. The real question is why they seem to interest Chernow so much, and—to borrow Chernow's phrase, "one wonders" if his audience is more seriously interested in marital fidelity than military history.

The real problem in Chernow's biography is that he's so comfortable in the world-as-it-is in the 2020s, that he doesn't want to make the effort of realizing that Grant's world was not like ours.

Chernow says that "For a young man, Grant was remarkably clear-headed and self-possessed in combat." That happens to be true. In fact there were times, like the prelude to the Shiloh debacle, when Grant was a little too phlegmatic for his army's own good. Still, there's no doubt his confidence and calm was a huge improvement on the various psychopathologies of most Federal commanders, at least those in the eastern theatre.

But even here, Chernow manages to give the impression that he can't really imagine the world of Victorian Anglos. Those people were terrifying. Bravery and calm in battle were the norm. Strachey was right to call them "demons," whether enlisted on the side of the angels, devils, or simply their own dedication to imposing their vast egos on the world.

Why, then, preface this proper praise for Grant's calm in battle with "For a young man..."? Does Chernow imagine that most young Anglo men of Grant's generation became hysterical in battle? Civil War officers skewed very young, much younger than the actors who played them in the 1993 film *Gettysburg*.

And an insane level of calm ferocity was the norm for them, especially those relatively elite kids who made up the officer corps on both sides. Matthew Broderick's fussy, pencil-necked, but heroic Col. Shaw in *Glory* is a closer approximation of a Victorian Anglo officer than

Chernow's lazy notion of Grant as a 21st-century centrist dad (and I frickin' HATE Matthew Broderick).

Chernow simply makes no attempt to enter the minds of these aliens—and they were aliens in ways that just don't interest him. He just doesn't see it as his job to reconstruct their terrifying mental world. His Grant is a cuddly and familiar one, which ends up being a diminution of his subject...at least for me. Maybe his audience wouldn't like a story about real Victorian officers. I'm pretty sure they wouldn't.

The Mexican-American War of 1848 would be a great opening for a revisionist historian. It was an ugly war, championed by slave-owning bullies. Grant himself saw this and denounced the war although he had participated in it as a young officer. You might expect a new biography of Grant to paint a much darker, more Goya-like version of the 1848 Mexican War. Well, once again, you'd be overestimating the alleged liberalism of Chernow and his readers. Chernow writes about the occupation of Mexico City like David Frum describing Baghdad circa 2005:

> "At first Grant witnessed brutal reprisals by Mexicans against their peers who had cooperated with the Americans.... But in time Grant saw how a wise, charitable policy toward a conquered civilian population restored peaceful conditions with impressive speed."

Only after a page of describing this enlightened occupation does Chernow feel obliged to add that "Other accounts of the American occupation depicted atrocities raging on both sides."

Yeah. Sorry Ron, but I'm gonna go with those "other accounts" you dismiss in a sentence. Victorian Anglos had very, very different ways of making war depending on whether the enemy were fellow white, Protestant English-speakers or those who failed to tick even one of those boxes.

Grant was more honest about this in his *Memoirs*, saying that he was far more horrified by the conduct of the Anglo Texans who were his ostensible allies than the Mexicans fighting to defend their country. With a non-white, Papist, Latinate-speaking population at their mercy, you can be very sure that the U.S. troops of 1848 raped and bayoneted their way through Mexico, even if they were too prim to write home about it.

Here again then is a disquieting glimpse of Chernow's strange demographic, obsessed with Grant's temperance and fidelity as a husband but uninterested in his conduct in a war which Grant himself painted very darkly. It is as if the lights go up in your mind and you see clearly the audience, which has given a standing ovation to the musical *Hamilton*. Always, by the end of the performance, "They think that things are all right."

Chernow's job seems to be building a more stable plinth for Grant's statue in the American pantheon. This plinth (and let's take a moment to enjoy that word. Life is not all Jeremiads, after all. Yea, there is yet time to savor the word "plinth" as we kick the mighty!) is based, not on any new insights, but on sheer bulk.

To fill the big plinth, Chernow relies, with his usual credulity, on Grant's many, many Parson Weems postwar hagiographers.[2]

Grant was the nation's darling for decades after 1865. And great generals generate legions of naïve fans. Chernow, astonishingly, has put together this biography not from primary sources but from the adoring biographies of Grant's fans.

Victorian Anglos were big on hero-worship (people bid huge sums for a vial of Garibaldi's bath water when the Italian hero visited London) and half Grant's former staffers published starry-eyed accounts of his greatness, which they all claimed to have discerned as soon as they saw him.

This is how every hero's story gets plinthed after the fact—but "plinthed" should not be the same as "pithed," and a lot of those guys wrote like pithed frogs. It's Chernow's job to sift the confetti and try to paste together the truth, rather than collect the confetti with a broom, convert it to pixels and offer it as proof that Grant, much like America, was great all along.

Because Grant WAS great. With all his grudges, his blind spots, his mistakes in the east and the carnage they wrought, his alcoholism and depression, Grant still stands as a figure worthy of awe. In fact, part of the awe he deserves derives from his murky, tormented Victorian mind. And that's why Chernow's Grant isn't just dishonest and Weems-like but deeply unfair to Grant. Chernow's Grant is boring, and Grant was not.

---

2. Parson Weems was an American author who wrote several biographies of historical figures, the most famous of which was his 1800 biography of George Washington.

If only Chernow had brought us a Strachey Grant. If only he had even a little interest in real military history. If only he grasped the notion that those people were not like us.

Or more selfishly, if only I hadn't wasted weeks slogging through the first two-thirds of this doorstop. No, I didn't finish it and don't plan to. Life, as they say, is too short. And this damned thing is way too long.

Chapter Fifteen

The Home Front

*W*omen's Civil War journals are remarkably interesting and surprisingly little studied until recently. Here I try to introduce a few of them but remember, these are just a few.

Women's Civil War Diaries, for the most part, don't describe the great battles. There are plenty of soldiers' journals that do that in a variety of tones and perspectives. Most women spent the Civil War at home, though they often fumed that they'd have loved to take up the gun and fight alongside their male relatives. Although several hundred of them did did dress as men and join the fighting, most settled into their assigned roles: feeding the family and livestock; keeping some sort of control over the household in their husbands' absence; and (most interestingly) maintaining a written record of events affecting the family during the war.

Recording such events was a woman's proper role in Victorian Anglo families. Since the Confederacy did not generally fund public schools, writing fluently was usually a skill confined to wealthy, educated women. Poor families had to settle for brief notes of births and deaths in the frontispiece of the family Bible. More privileged women kept voluminous records of the war years, and published them after the war, often to rehabilitate Confederate "heroes" as part of their family-defender role.

Some of these wartime journals became bestsellers, some languished in drawers and attics until tempers had cooled. Some were heavily revised for a postwar audience, though passed off as unaltered war-time records.

In this article, I'll talk about six women's wartime journals:

A Diary from Dixie by Mary Boykin Chesnut
A Union Woman in Civil War Kentucky by Frances D. Peter
Another Year Finds Me in Texas by Lucy Pier Stevens
A Confederate Girl's Diary by Sarah Morgan Dawson
Brokenburn: The Journal of Kate Stone
A New England Woman's Diary in Dixie in 1865 by Mary Ames

1) *A Diary from Dixie* by Mary Boykin Chesnut

The best-known Civil War journal, *A Diary from Dixie* by Mary Boykin Chesnut, is one of the best, but also one of the least "authentic." It passed as a truthful record of one Charleston woman's reminiscences of the war, until a hard-nosed critic pointed out that Chesnut, who was a very sophisticated product of the Charleston elite, had demonstrably revised her manuscript to make it more acceptable to a postwar audience.

Chesnut never admitted she'd done a thorough editorial revision, and her fans, who were enthralled with the Lost Cause narrative, were enraged that anyone dared to question the provenance of her text. In 1981, Kenneth Lynn called her diary a "hoax" because it had been heavily revised.

He was answered by William Styron, then a very popular novelist, who took the role of Chesnut's defender.

Both positions in this debate have aged badly. It seems naïve to suppose that Boykin Chesnut never changed anything before publishing her diary, and even more naïve to condemn the entire book as false because it had been revised.

It's impossible to divide memoirs into sincere/truthful and false/fictional. Rousseau claimed to have written the only totally truthful memoir, but the notion of a totally truthful Rousseau is risible on its face.

All memoirists start with bits of truth, simply because the details of one's life that one remembers are too good to waste; and all memoirists excise or add material. Untruth does not consist only in telling outright lies but in excising some memories and amplifying others. Mark Twain himself has Huck Finn make a more sophisticated summary of the matter, describing his own edits of his childhood memories:

> "That book was made by Mr. Mark Twain, and he told the truth, mainly. There was things which he stretched, but mainly he told the truth. That is nothing. I never seen anybody but lied one time or another, without it was Aunt Polly, or the widow, or maybe Mary. Aunt Polly—Tom's Aunt Polly, she is—and Mary, and the Widow Douglas is all told about in that book, which is mostly a true book, with some stretchers, as I said before." [1]

---

1. *Huckleberry Finn: Tom Sawyer's Comrade* (Charles L. Webster and Company, 1885) p.2.

Note that Huck Finn, a chivalrous soul, exempts all the "respectable" women in his world from the charge of "stretching" the truth. This is a matter of provincial courtesy; accusing respectable women of lying is an outrage.

But it's clear that even Huck is exempting these women more as a matter of respect for one's betters than in the real conviction that they never lied. Twain even wrote many times of the need for lying in polite society, most directly in his short story, "Was It Heaven—Or Hell?," in which two upright old women must face the fact that lying may be the only alternative to outright cruelty.

Chesnut spent years editing the journals for publication, and it would be absurd to think that she never once yielded to the temptation to change her record of antebellum Charleston to suit a postwar — and most importantly post-Emancipation, audience. Remember, truth means not cutting anything you ever wrote and it's very very unlikely that Chesnut never cut an embarrassing passage.

Some key scenes seem to me (and I can't prove it) wholly fictional, inserted to portray Chesnut as compassionate toward the enslaved. Above all, the scene in which MBC sees an attractive slave woman being auctioned to an audience of lustful men seems to be a post-Emancipation edition:

> "I have seen a negro woman sold on the block at auction...I was walking and felt so faint, seasick. The creature looked so like my good little Nancy, a bright mulatto with a pleasant face."

Chesnut identifies briefly with the slave woman, interpreting her expressions:

"She was magnificently gotten up in silks and satins. She seemed delighted with it all, sometimes ogling the bidders, sometimes looking quiet, coy, and modest, but her mouth never relaxed from its expanded grin of excitement. I daresay the poor thing knew who would buy her."

Chesnut then describes her own reaction to this tableau:

"I sat down on a stool in the shop and disciplined my wild thoughts...[telling myself] you know how women are sold in marriage from queens downward, eh? You know what the Bible says about slavery and marriage; poor women! Poor slaves!"

This passage, smoothly eliding the difference between a Planter woman and a slave being sold, culminates in the phrase "poor slaves!" which the reader is meant to understand as applying equally to the enslaved woman on the block and Chesnut herself. It is a ridiculous postdated epiphany, clearly postwar and intended to use high-Victorian loathing for the specifically sexual side of slavery to cast Chesnut and her readers as innocent, sympathetic bystanders in the spectacle of sexualized slavery.

Chesnut's wartime diary may have included a mention of the woman being auctioned, and may have ended with an expression of loathing for the male audience, but this page-long bathos, inserted early in the postwar diary meant for publication, seems to me to be an amplification of what was originally a simple diary entry.

This is very postwar prose. Boykin does not normally refer to the enslaved in prewar Charleston as "negro women." They are referred to by their first names ("my Nancy" for example) or their position in the household hierarchies.

Boykin was a genuine intellectual in a circle that valued "accomplishments" like speaking fluent French (as she did) but preferred decorum to argument among elite women. Wit was valued, but a Planter woman's most basic role was to have many legitimate descendants. When Chesnut quotes her father-in-law praising his wife, it is for her fecundity:

> "Wife, you must feel that you have not been use-
> less in your day and generation. You have now 27
> great-grandchildren."

Mary Boykin Chesnut, who is 38 years old at the start of the war and a childless anomaly in a wildly fertile milieu, is "useless" in her father-in-law's terms. He may indeed have meant to accuse her by way of praising his wife. At any rate, Mary lives with the stigma of childlessness and it contributes to her bitter tone toward her peers. She writes like someone ready to take offense at any sign of a slight and her gloomy predictions about the fate of the Confederacy are generally proved correct.

But she is not immune to fatuous sentimentality when judging Confederate commanders. Above all, the diary is full of the ridiculous cult of John Bell Hood, one of the worst Confederate generals (I'd say the very worst of all), but a "knightly" hero for the Planters and, above all, Planter women. The bathos begins the moment Mary introduces him in her diary:

"The famous colonel of the 4th Texas, by name John Bell Hood, is here—him we call Sam, because his classmates at West Point did so...When Hood came with his sad Quixote face, the face of an old Crusader, who believed in his cause, his cross, and his crown, we were not prepared for such a man as the beau-ideal of the wild Texans. He is tall, thin, and shy; has blue eyes and light hair; a tawny beard, and a vast amount of it covering the lower part of his face, the whole appearance that of awkward strength...the fierce light of Hood's eyes I can never forget."

The "Crusader" and his new fan meet many times after that:

"Hood came to ask us to a picnic next day. We were to have bands of music and dances...Hood and his staff finally came galloping up, dismounted, and joined us. Mary Preston gave him a bouquet. Thereupon he unwrapped a Bible, which he carried in his pocket. He said his mother had given it to him. He pressed a flower in it."

Hood doesn't flirt with Mrs. Chesnut for her witty sallies. His war wounds were often props for his assiduous courting of influential Confederates, from Jefferson Davis down to the loathsome General Bragg. And his lobbying was on behalf of one of the very worst military minds in the Confederate elite, himself.

Hood knew only one trick: attack, attack, and attack again. It had worked for R. E. Lee in the Seven Days Battles, but that was because Lee was facing McClellan, who didn't have enough nerve to make a left turn on a green light. McClellan, whom Lee had known (and identified as a coward) at West Point, cracked and ran under constant attack. Other Union generals did not. They had learned by 1863 that any well-drilled troops would hold if they were well entrenched, and would inflict such horrific casualties on their attackers that repeated attacks would destroy the attacking force.

Hood never learned this, as he showed in the debacles at Ezra Church, Franklin, and Nashville, and in his policy of defending the Deep South by endlessly attacking entrenched positions he single-handedly destroyed the Confederacy's second-largest army. And Mary Chesnut's circle in Charleston never figured it out either, as is shown in their awe for Hood's maimed body and their disdain for his rival Joe Johnston's strategy of attrition and slow withdrawal.

So it's not as if Mary Chesnut is immune to faddish hero worship. Her alienation from the fecund elite of the Confederacy does not save her from admiring the very worst of its commanders.

The tone of Chesnut's diary changes in the final chapters, showing (probably) revisions to suit the prevailing bathos of the Lost-Cause narrative that was becoming orthodoxy in the South and even in left-wing Northern circles. (For proof of that claim see Edmund Wilson's bizarre attitudes toward the Union and the Confederacy in *Patriotic Gore*, published in 1962. By way of rejecting the narrative of the capitalist North, Wilson swallows whole the silly Agrarian lies of the antebellum South.)

For Mary Chesnut, a much colder mind, the shift to bathos comes harder. Her final scenes show the former enslaved as simultaneously loyal and affectionate to their former masters but also superfluous.

"The fidelity of the negroes is the principal topic" in the postwar South, she says; but she turns in an instant to contempt for them: "The negroes would be a good riddance. A hired man would be a good deal cheaper..."

She retells all the usual tales of verbal triumphs over the conquerors, as if they provided real consolation: Planters who saved their real gold by pretending it was mere plate, so that the Yankee plunderers give it back, saying, "You don't fool me that way; here's your old brass thing; fork over the silver!"

The final entry, clearly chosen to set the tone of the highly edited text, gives a set of cues to future protectors of the Lost-Cause narrative:

"August 2 [1865]—Dr Boykin and John Wither-
spoon were talking of a nation in mourning, of blood
poured out like rain on the battlefields—for what?"

This might seem like a valid question, but the unspecified inter-locutor responds,

"Never let me hear that the blood of the brave has
been shed in vain! No; it sends a cry down through all
time."

This is the point of the whole diary: To offer a stance that would be maintained and defended throughout the twentieth century, only giving way to pressure in the 21st. It demands silence on the appalling casualties of the lost cause ("Never let me hear that the blood of the brave...") and a commitment to transmitting the Lost Cause's "cry" "down through all time."

Astonishingly, this content-less and irrational "cry" was indeed maintained, and immortalized in monuments, for more than a century—which, compared to most genuinely heroic past events, is quite a long time to be commemorated—even though no one ever specified with any clarity what the "cry" meant.

2) *A Union Woman in Civil War Kentucky* by Frances D. Peter

Are there women's wartime diaries that can be taken as "real" narratives, written in haste, without calculation? Not really, unless one accepts the view that humans can only show cunning via revision. Many, many people think faster than that. If you've ever needed to lie (and you have), you will recall that your invention worked at light-speed, revising sentences microseconds before saying them.

But we can say that some diaries were uncorrected first drafts — because their authors died before they got the chance to labor over detailed revisions.

Frances Peter, one of the greatest of Civil War diarists, died at 20, in August 1864, before she had the so-called "maturity" to modulate her fierce and merciless intellect. She had been writing her diary since the age of ten, knowing she was probably going to die very young.

Peter, who recorded the war as it happened in her hometown of Lexington KY, was one of 11 children in a strong Unionist family, coddled and cherished by her family for her sufferings and her dwarfism and illness. Like the writer she resembles, Flannery O'Connor, she seems to have sensed that she had very little time and written her impressions of wartime Kentucky in letters of fire.

Frances never mentions her illness or disability. She focuses on what she hears, reads, and can see out of her window. Lexington was a contested city in the early war years, and Frances writes in a fine rage about how a local adolescent, "Morgan" from the hated Morgan clan, raises a secession flag over Transylvania College, where her father teaches:

> "Last evening Key Morgan, with two or three oth-
> er boys, went to the janitor of Transylvania College
> and got the key to the door leading onto the roof on
> the pretext that a ball had been thrown up there and
> hoisted a Secession flag up on the college. The janitor
> saw it and cut it down."

"Morgan" is a powerful name in Lexington. The Confederate hero of the region was John Hunt Morgan, a rich boy who played with organizing private armies in Kentucky well before the war. He attended Transylvania College, of course, just like every single future Confederate officer west of the Mountains, then jumped at the chance to organize an independent cavalry force, ostensibly on the Confederate side but really operating more like the Free Companies of the Thirty Years War. Confederate officials came to despise these freebooters even more than Union officers, considering them worthless in battle and only good for battening on their own people, stealing their livestock in the name of military necessity.

The Morgan family lived one block away from Frances's home and were a typically large frontier family, and most of them fought for the Confederacy. They were related, by blood or marriage, to other Confederate figures like A.P. Hill. Basil Duke, another Morgan connection, was called the "heart and soul" of Morgan's force.

Frances, confined to her house by her deformity, stares toward the Morgan house with unwavering hostility. The Morgan house becomes an evil presence in her diary. She imagines it as dominated by John

Morgan's mother Henrietta, who survived the war (unlike most of her sons) and of course Transylvania College, the center of town society, takes center stage as a sort of Minas Tirith to the Dark Tower of the Morgans' house.

There's a constant tension between the College, which was her father's workplace, and the Morgan house, always threatening violence.

This was early in the war, when both Unionists and Secessionists in Lexington were in a relatively tolerant mood. The man who planted the flag, Key Morgan, was not punished. This is unusual in a civil war. Most civil wars start with civilian massacres and get even more cruel as they go. Key Morgan was lucky to do his little prank early in the war. Later on, he might have been stood up against a wall and the family house burned.

Morgan's cavalry did well through 1862, due largely to the passive (if not traitorous) policies of Don Carlos Buell, who seemed to be more sympathetic to Confederate guerrillas than to his own Union volunteers. At this stage of the war Morgan's men could ride openly through Lexington. Frances looks out at them with a hygiene-based contempt that echoes the rhetoric of Matty Ross in *True Grit*:

> "[A] nasty, dirty looking set they were. Wore no uniform but were dressed in gray and butternut jeans and anything else they could pick up, but were not quite so dirty looking as Kirby Smiths. They looked like the rag-tag and bob-tail of the earth, and as if they hadn't been near water since Fort Sumpter fell."

What makes Frances's account of Morgan's comings and goings so vivid is that she keeps track, in the way that many housebound people do, of her enemy's movements. When Morgan comes back from his raiding to visit the family home, Frances is watching angrily:

> "They passed along the side street by Mrs. Morgan's, the officers dressed in gray or black, the officers wearing different kinds of flat hats and feathers with cockades or streamers. The men in clothes of various colors, only uniform in respect to dirt."

That's a wonderful line, "only uniform in respect to dirt," playing off the root meaning of "uniform" ("the same") to suggest that filth is the badge of Morgan's men in itself. And she wasn't exaggerating much, about either the flat hats or the grime.

> "Some of Morgan's men in a Union prison. They appear to consist entirely of hats, boots, and beards...but even so, I would not care to offend them needlessly. Or even needfully."

This snobbery about one's enemies occurs in many women's CW diaries. When Mary Chesnut wants an anecdote that will convey her contempt for Lincoln's family, she repeats a story about Mrs. Lincoln's alleged trashy penny-pinching:

> "I hear that Mrs Lincoln means to economize. She informed her major domo that they were poor and hoped to save 12 out of their $20,000 salary..."

Though her milieu is more modest than Chesnut's (almost every-body's was), Frances takes her revenge on the invaders exactly the way that Confederate women do: By sneering at their manners, their garb, their ignorance of local customs.

Because women, or at least middle-and upper-class women, don't expect to be shot or beaten by the enemy, they take every opportuni-ty of humiliating the invader. Frances brags that Lexington's Union ladies strung a rope across the road by which Morgan's men will enter the city, and tied a big American flag from it, so the Confederates will have to pass under it. She rages at the sight of Morgan's men violating her town.

> "How mad we were to see those rascals! We could hardly keep within bounds. I heard say [sic] that a young lady...said to her sister, 'Well I can't stand this any longer, I must say what I think.' 'Look here," said she to one of the [Confederates], I say 'Hurrah for Lincoln. I mean the president of the United States if you don't know. And hurra for the Union and the Union soldiers.' 'You do?' said the man getting very angry. 'Yes, and hurra for Hell and damnation.'"

These stories appear in every CW woman's diary I've read. They show a real difference between the US Civil War and most rebellions. Within certain limits, outside of Kansas/Missouri, and especially in the first phase of the war, most regular soldiers avoided shooting white civilians, especially white women of the "respectable" classes. That may not seem like much, but it's truly unusual and worth noting.

Of course, women didn't know that ahead of time, so going up to an enemy soldier and insulting his cause was very risky. If it happened, that is. These stories of defiance are almost always second-hand; "a friend," or "a friend of a friend" got in the enemy soldier's face. They're very like the male adolescent stories of fights that almost happened; they stop short of real violence and focus on verbal one-upmanship.

The brief anecdote which follows Frances's "Hurrah for Lincoln" story would probably have wounded Morgan's troopers much more deeply. She describes the Secesh ladies' disdain for the dirty, white-trash Confederate troopers:

> "The secesh ladies didn't seem to like their looks any
> more than we did... When the common men came to
> their houses to ask for something to eat they had them
> taken in at the back gate to the kitchen and let the
> negroes wait on them."

Frances never got the chance to celebrate the Union's final victory. She died on August 5, 1864, after an epileptic fit which she and her family had probably been expecting and dreading for years.

But while life stirred in her, she worked out her anger in patriotic invective. In this way too, she reminds me of Flannery O'Connor, taking revenge in advance on the slack-souled civilians who will inevitably outlive her, thanks to their lukewarm hearts.

Frances's world begins the war in a rather lukewarm fashion, especially on the Union side. In most Southern towns, Unionists were generally middle-class professionals or shopkeepers who just wanted to keep their heads down and make money. (It was different in the Appalachian districts of Eastern Kentucky.) Most urban Unionists in the border states were initially moderate, simply because they didn't want to be stabbed or shot or hanged or burned out of their homes.

They had read for years of the border wars to the west, in Kansas and Missouri, where the Secesh always seemed to strike first and hardest. They wanted to keep a few humane norms in place as long as possible. So in the first two years of war, by Frances's accounts, there's still a certain reluctance to execute prisoners. Even pro-Confederate Kentuckians are slow to kill their Unionist neighbors...at first. Noone but a few 20-something crazies wanted Lexington to be the next Lawrence, KS. So, in some of Frances's stories from the early war years, there's a comic reluctance to kill prisoners, very unusual in the records of guerrilla warfare:

> "A Union civilian was stopped outside Nashville by a party of rebels. He told them he was a shoemaker and was going through the country buying leather. They asked where he got that [Union] uniform and saddle he had in his bag and wanted him to take the oath to the confederacy. This he refused and said he couldn't tell who the uniform etc. belonged to. Then they wanted him to take a parole, but he wouldn't. So they asked him not to give aid or comfort to the enemy or reveal what he saw that night, which he agreed to."

This anecdote is almost comic, as the guerrillas are bargained down to a one-night vow of silence. If this had happened in Kansas or Missouri rather than early-war Kentucky, there would have been a different ending, with a body left by the road. This might be because every Lexington family knows every other family, even the Morgans. After all, even the Morgan family has its weak link, as Frances tells it:

> "Captain Charleton Morgan was sent off to a POW camp. He cried like a child. He has always had the reputation of being weak-minded and childish. Mrs. Morgan had a great many visitors this evening, friends and relatives come to console about the mishap and Lincoln's barbarities I suppose."

Charleton Morgan seems to have been one of the weakest links in a very large family, and the thought of the runt of the litter in Yankee hands may have led to Morgan's initial reluctance to kill captives. This squeamishness wore off all too soon for many guerrillas, but in 1862, at least in Lexington, it seemed to be the rule. Morgan never had the grim reputation of other Confederate guerrilla leaders like Quantrill or Anderson, but Frances's stories about them show fear as much as hostility. The Morgans were not by any means "white trash," but they were the sort of smalltown Southern gentry who were not averse to sudden ultraviolence.

Frances's family were genteel in a more northern or European pattern, one which didn't place as much stock on individual proficiency with weapons. Her hostility to the insurgents down the street is tempered, accordingly, with wariness about their hair-trigger pride, in which gentility has no resemblance to the current meaning of "gentle" but rather descends directly from the Elizabethan model: "Make not too rash a trial of him, for/He is gentle and not fearful" —in antebellum Dixie terms, "He's got a middle name that goes back to 17th c.

Virginia, and will therefore gut you like a fish if you even look at him
the wrong way."

The Unionists in border towns were always playing catchup against
the reflexive violence of the local elites. They took to killing their
neighbors more slowly, and when they did it was with more delibera-
tion. But guerrilla wars always involve loss of discipline, as small units
have to make their own decisions with the memory of past grudges
driving them to cruelty as the war continues. In 1862, at least, Frances
tells the story of a "rebel" spy caught with overwhelming evidence, yet
not (as far as she tells the reader), executed:

> "One of Morgan's spies, a scoundrel named Bill
> Owens, was taken up today disguised in women's
> clothes. The aforesaid courier was in bed when the
> house was surrounded. A number of papers and let-
> ters were found on him."

Perhaps he was hanged, and Frances simply doesn't want to say so.
Frankly that's not the Frances I came to know and love while reading
her journal — that Frances would have gloried, in the true Flannery
O'Connor spirit, in the sight of gallows decorated with Confederate
"scoundrels" —but it may be that, since her information about distant
events like the taking of a spy only comes through relatives' stories,
they wanted to spare her the grisly details. If so, they were wasting their
time.

Frances records the capture of a female Confederate spy, with no
sign of compassion:

"The notorious rebel emissary Miss Moon was captured the other day and a large number of letters and
papers, bottles of quinine and morphine were found
on her. She has made several trips south for the secesh."

Once again there's no mention of "Miss Moon's" fate, nor any sign
that Frances took pity on the courier for trying to bring pain relief to
wounded soldiers. Frances has no pity for "secesh ladies":

"Five secesh ladies were arrested this evening for ex-
pressing treasonable sentiments and sentenced to be
sent south to their dear rebel friends. They said they
preferred staying here. They were told they must either go south or to jail, and as they refused to go
to Dixie, they were put in jail. After a short while
they became quite tractable and took the oath of allegiance."

You can hear the scorn in her account of these weak reeds, whose
allegiance to their cause failed at the mere prospect of exile. You get the
feeling that Frances, who probably knew her life would be short and
painful, had real contempt for these "secesh ladies" feeble allegiance
to their "dear rebel friends." There is no fervor like that of a pubescent
outcast contemplating the unaccountable weakness of physically perfect people who quail in the face of mere exile — not even death, just
having to leave home!

In her entries from 1864, the last year of her brief life, Frances saw or
heard of all the brutality even she wanted. When Morgan's men make

another raid on Lexington in the course of their great raid of 1863, she describes a much grimmer scene:

> "On the march three [Union hostages] fell with exhaustion. One of them was dragged before the artillery, which passed over him, crushing him to death. The other two had their brains knocked out with muskets."

This may be just a gory story Frances heard from her relatives, but it reflects the fact that by the summer of 1863, chivalry had disappeared from the war, especially the war between Union and Confederate cavalry in the border states.

The one issue on which white Unionists and Secessionists agree is, alas, "the negroes." Kentucky Unionists had made their feelings clear by resigning in large numbers when Lincoln signed the Proclamation, and Frances shares their aversion to racial equality:

> "[Union General Gilmore has issued an order that no negroes are to be allowed to enter the camps and that any negro found in the United States uniform shall be arrested and punished, which I think is very right..."

Frances repeats the second-hand justifications for this racist policy, as she gets them from Kentucky newspapers

> "The negro regiments...have proved a failure, the negroes refusing to work and shirking... So say some of the papers, and others make out that everything is

going well and the black soldiers are so patriotic and
etc. I doubt it... I think it is acting against the Constitution to make soldiers of the blacks, and however
much the abolitionists may say to the contrary, they
will find that this arming and equipping [sic] of the
negro regiments is a mere waste of money..."

Frances remains fiercely Unionist as long as the conflict is between
white Unionists and Secessionists. In fact, in line with the usual course
of civil wars, her position, and that of her community, hardens as
the war drags on. The Morgans and the Peters may have been bitter
enemies earlier in the war, but there was a superficial détente between
them as they passed each other in the street. Frances contents herself,
at this stage with a little malicious gossip, some of which is genuinely
funny:

"Mrs. John Morgan['s] letters that have been intercepted [say that she had] a 'bewitching bonnet which
my noble husband brought me when he came back
from his last raid.' But [she says] she was almost out
of shoes, she couldn't get more until her 'noble husband went on another raid.' No doubt John [Morgan]
brought her a great deal of stolen finery and expected
to take her no end of pretty things but if she don't get
any shoes until John brings them I am afraid she will
be in a bad way."[2]

---

2. Because Morgan had been captured by Union forces on June 26,
   1863, a few days before Frances writes this journal entry.

But as Kentucky comes under enduring Union occupation in 1863, open displays of rebel sympathies are no longer permitted. As always, it is the middle- and upper-class women of the dissident (Confederate) community who dare make public displays of their dissent, and eventually they pay for it:

> "Saturday May 23 1863: An amusing incident happened in Louisville the other night. At the theater during the intermission between the plays the band played some national airs [patriotic Union songs], on which a number of ladies got up and flirted [sic] out of the room. The next night the manager came on the stage and announced that the band was about to play some national airs and all those who were too much opposed to the government to listen to them had now an opportunity to leave. As on the previous night, a number of ladies got up and flirted out of the room but at the door they were met by the Provost Guard, who marched them off to jail."

Ah, a night in jail—the ultimate punchline!

By the standards of many insurgencies, a night in stir is a very mild punishment, but one does get the sense that it was lucky for many border-state hotheads that the war ended in 1865. By that time, too many had died and tempers were growing short even for the phlegmatic Unionists. A few days after writing up the arrest of the Secesh ladies, Frances notes General Burnside's banning of two Northern newspapers, the Chicago Times and New York World, for seditious articles.

The Confederates had purged their border communities by quick pogroms at the beginning of the war (or earlier.) The Unionists were characteristically slower and more thorough. They were gentle and slow to anger, but after the Union victories at Vicksburg and Gettysburg in July 1863, it came to the same thing.

Frances monitors any deviation from strict Unionism in her jurisdiction, the world of middle-class Lexington. She is particularly alert for lukewarm sermons from the local pulpits, as she notes in this diary entry from August 6, 1863, just a month after Gettysburg:

> "Thanksgiving day passed very quietly... Mr. Brank [Pastor of] 2nd Presby[terian Church] made a pitiful excuse about being afraid of dividing his congregation if he had service today, and some of the other [closed] churches have secesh ministers."

Frances would have made an excellent NKVD operative. In fact, America might have been better off if a little NKVD-ish rigor had been imposed on the defeated Confederacy.

But Frances died in, if not of, the war. She had the fatal epileptic seizure months before the end of the war. It's a shame. She would have enjoyed the Victory celebrations, especially the sullen silence from the Morgan house a few blocks away.

***

Chesnut and Peter are the two greatest female diarists of the Civil War—Peter because she's an under-appreciated observer and stylist, and Chesnut because she made herself the tragic heroine of the Lost Cause—but many other women kept wartime diaries, aware they were living through a heroic era. I'm not aware of any other diaries as good

as the Big Two, but there are many that highlight different aspects
of women's Civil War experience, from burial customs to romance,
(which, by the way, are not such disparate topics in the Victorian
mind).

Here are a few more I've read.

3) *Another Year Finds Me in Texas*, by Lucy Pier Stevens

If you're looking for corpse-porn, look no further than this Mid-
western woman's diary of a sojourn in Texas. Which is odd, because
the author was nowhere near the war. Daily life in a Texas town gave
her enough material. Life in the mid-nineteenth century was full of
slow death, and Victorian culture was not at all shy about recording
the details. It's a strange experience for 21st c. readers, encountering
a culture which is prudish, sometimes comically prudish, about anything involving sex, but eager to share the XXX-rated details of death
and decomposition.

Most of Stevens's anecdotes open with a description of whatever
illness is afflicting a relative or friend. She then records the death,
focusing on the grief of close relatives:

> "Found Mrs.—in the deepest grief. Poor, poor lone
> woman. Would that I could be a comfort to you, but
> no one could ever fill the place of your affectionate
> daughter."

The next day's entry is grimmer. August 1864. Galveston is hot and
humid, the body is decaying:

"The body looked as natural as it could be in the morning. But at half-past ten she commenced changing, changed so very fast we feared her eyes would burst before the coffin would come, but happily they did not."

Everyone spends some time in bed being sick. Many of them just die. Another diarist quoted by Stevens says:

"It seems to me that it is nothing but sickness and death now days. Ella is the eighth person I have been with when they died within five months. I am tired, tired, so tired."

Stevens herself gets sick:

"As anticipated I had a severe spell of sickness. The chill I had was congested and I have had congestive fever until a few days ago. This is the third time in my life I have not been expected to live."

There are no effective medicines in Stevens's world. With the Confederate coasts blockaded, not even palliatives like quinine and opium are available.

4) *A Confederate Girl's Diary*, Sarah Morgan Dawson

Dawson is more the orator. She makes speeches even to her own diary:

"Ah, Liberty! What a humbug. I would rather belong
to England or France than to the north. Bondage,
woman that I am, I can never stand. Even now the
northern papers distributed among us taunt us with
our subjection! Ah, truly this is the bitterness of slavery... [kinda a lack of self-awareness you might say]
to be insulted and reviled by cowards who are safe at
home!"

Her concern, which she shares with some of the other diarists, is
that in her culture you're supposed to be married by 24, some say 22,
and that date approaches. All the men are in the Confederate Army
and they only visit occasionally.

"If I could so far forget my dignity and my father's
name that I could cultivate the notice of gentlemen—!!"

Toward the end of the war she and her friends resort to Divining,
supposedly as a joke but not entirely.

"Six of us around a small table invoke the spir-
its...Our first question was how long before peace?
Nine months was written. Which foreign nation
would recognize us first? France, then England in
eight months. Who was Miriam to marry? The captain of a battery. Of course I do not actually believe in
spiritualism, but there is certainly something in it."

Chapter Sixteen

Checking in on the Lost Cause, 156 Years Later

*T*here's always been a second school of pop-Civil War history, much more aggressive than Chernow's general optimism. These are the pro-Confederate "Lost Cause" writers and, amazingly enough, they're still around.

American Conservative is a strange magazine. They publish some great articles—why, I've published there myself, a sure proof of their good taste. They have a fine record for exposing the expensive debacles that pass for American foreign policy. But when it comes to domestic

American topics, they go and print stuff like Helen Andrew's "Reconstruction Revisionism."[1]

I'm not saying a site has to agree with me all the damn time. I hate that kind of talk. But this, AmCon, this is just bad writing. So bad, in fact, that I don't want to give Andrews' nonsense more focus than it deserves. I'm sure lots of other people have done thorough demolition jobs on it. I'm a latecomer to the boot-party because I was doing real work on the Radio War Nerd Civil War series. In fact, some of the Civil War memoirs I've been reading refute Andrews's thesis directly.

But that's the thing here: the article is useful as an example of a kind of rhetoric that's very powerful with some audiences. And since this rhetoric is the pure product of the thesis-driven Reading & Composition courses I used to teach, my reaction on reading Andrews's article was remembering all those eager would-be contrarians whose whole writing style depended on not knowing a thing about the topic, but knowing all too well a few basic rhetorical moves.

Their method works very well, even if it isn't very impressive. Reminds me of something a friend of mine who gets in fights once said to me: "The fact is, the ol' one-two works just fine 99 times out of 100."

And the best thing about this Ignorance-Fu is that it works better if you don't know a thing about the topic. This is the ultimate in what Aristotle called "a portable faculty."

Andrews's capsule bio, attached to her article, seems to confirm that she has bravely avoided learning anything about the US Civil War or its aftermath, preferring to keep the ideological high ground rather than charge into the abatis of mere fact:

---

1. "Reconstruction Revisionism" by Helen Andrews for American Conservative (December 11, 2021)

"Helen Andrews is a senior editor at The American Conservative, and the author of BOOMERS: The Men and Women Who Promised Freedom and Delivered Disaster (Sentinel, January 2011). She has worked at the Washington Examiner and National Review, and as a think tank researcher at the Centre for Independent Studies in Sydney, Australia. She holds a Bachelor of Arts in Religious Studies from Yale University. Her work has appeared in the New York Times, The Wall Street Journal, First Things, The Claremont Review of Books, Hedgehog Review, and many others."

There's nothing in that bio to suggest that Andrews has wasted even a minute of her busy life on the Civil War or Reconstruction. But then, that's the beauty of the undergraduate martial art she exemplifies: the less you know about a subject, the more boldly you can make the worse cause seem the better.

This is the fighting art that has made many a dweeb who hit his intellectual peak in the SAT into a wealthy darling of one or another online ideology. In that way, and only in that way, it's worth looking more closely at Andrews' work.

Granted, "Reconstruction Revisionism" is much longer than five paragraphs. But be ye not disheartened, because this is nothing but a five-paragraph Reading & Comp product, of the sort that made Ben Shapiro a rich man. The article's length is simply de copia, or as civilians call it, padding. Rip off the padding and you get the same, simple but effective rhetorical tricks we talk about over and over on RWN.

Above all, the trope of starting date. This might be the most important move anyone arguing about history can make. When does the story begin? Choose your starting date wisely and the rest of the argument writes itself, especially when you're writing to an audience that knows nothing about the topic.

With that in mind, here's the first paragraph. Note that Andrews doesn't want to talk about the 19th century at all. Her starting date is "most of the 20th century.":

> "The wholesale reinterpretation of history around a left-wing narrative about race, which the 1619 Project is trying to accomplish for the rest of the American story, was first trialed on the history of Reconstruction. For most of the 20th century, Reconstruction was seen as a squalid and shameful coda to the Civil War when Northern Radicals and carpetbaggers enacted their wildest fantasies of humiliation and spoliation on a prostrate South. Starting in the 1960's, a group of revisionist historians began arguing that Reconstruction had actually been a noble experiment in interracial democracy, too quickly abandoned. It is noteworthy that this line started being touted only after the last people with firsthand memories of Reconstruction had died."

The rhetorical template for this paragraph is the Golden Age fallacy. Any time you read the word "anymore," you're probably encountering that fallacy. "Nobody does X anymore" is the usual form, and the simple response is "When did they?"

In Andrews's first sentence, you see this Golden Age trope sprawled across such a wide field that you already suspect she doesn't know or care about Reconstruction as an historical period at all. She starts with "reinterpretation of history," which is not only vague but tautological, since "history" is, by definition, a "reinterpretation." Something happens, and somebody "reinterprets" it. When was the golden age, the time when there was no "reinterpretation"?

We find out later in that first sentence. The "reinterpretation," Andrews says, was "...a left wing narrative about race," and it's linked to "the 1619 Project." This makes it pretty clear that Andrews's only interest is in contemporary American discourse—which in itself is proof of very bad taste, if nothing else.

As for the notion that "race" is a 21st-century obsession that has been imposed on the historiography of Reconstruction—well, that's so ridiculous it made me wonder if Andrews is really as ignorant as she sounds. Because honestly, how could anyone imagine that "race" was not an issue in Reconstruction? No one in the 19th century ever claimed that, least of all the opponents of Reconstruction. Their entire position was that Reconstruction would "elevate" the freed black people of the former Confederacy—which they considered a horrific prospect. They were ferocious on this point.

Imagine, if you will (Rod Serling voice here), Helen Andrews, time traveler, trying to explain to Nathan Bedford Forrest and the other founders of the Ku Klux Klan in the postwar south that a bunch of "left-wing" northerners are trying to make the narrative "about race." There would be a good deal of head-scratching followed by either a lynching (which would be a terrible shame, because if there's one thing the world needs, it's Religious Studies majors from Yale) or an earnest attempt to explain, "Ma'am, you don't seem to understand here! They're trying to make it NOT about race! They're trying to elevate the [n-word] and pretend he's just the same as us white people! You got it backwards!"

It will be a terrible shock for Ms. Andrews when she discovers that an obsession with "race" is not a purely 21st-century phenomenon. God forbid she should ever read the secession manifestos of the southern states in 1860 (South Carolina) and 1861.

She would be amazed at the extent to which the "narrative" was, for those people, "about race"! The terms "African slavery" and "negro slavery" are repeated endlessly in the seceding states' manifestos. The Confederates were not shy people. It was, to put it bluntly "about race." With a vengeance.

Andrews simply isn't interested in Reconstruction. Her only interest is now, now, now, and her antithesis is between the good ol' 20th century and the horrible leftist now. Reconstruction is used purely as an example of that antithesis:

> "For most of the 20th century, Reconstruction was seen as a squalid and shameful coda to the Civil War when Northern Radicals and carpetbaggers enacted their wildest fantasies of humiliation and spoliation on a prostrate South."

Andrews's focus on the 20th rather than the 19th century is the one thing in her article that makes me suspect she's dishonest rather than simply ignorant. Because it is quite true that for "most of the 20th century," especially the first half, there was a consensus—absurd and shameful, but real—that "Reconstruction was...a squalid and shameful coda to the Civil War." The problem was that the Civil War and Reconstruction happened in the 19th century. The twentieth century was indeed devoted to sinister, malign, very much right-wing reinterpretation of that history.

Reconstruction was a struggle about the role the freed slaves would play in the American polity. "Northern radicals" thought they should be treated like other American citizens. The defeated Confederates begged to disagree. Eventually, in a long and dismal struggle characterized by the lazy, self-serving compromises of the U.S. elite, the ex-Confederates were allowed to disenfranchise the freed slaves, reinstitute a system that was very much "about race"—that was, in fact, a return to race-driven serfdom for the black people of the south, but without outright "African slavery."

The failure of Reconstruction was that, as the Federal government became distracted and absorbed in other issues, the Planter elite made a deal with them on the lines of "Let us enserf the black people again and we'll reenter the U.S.A. that we openly despised until you conquered us." It was the dirtiest of deals, and recognized as such by many 19th-century commentators.

> "[The people] want a reconstruction such as will protect loyal men, black and white, in their persons and property; such a one as will cause Northern industry, Northern capital, and Northern civilization to flow into the South, and make a man from New England as much at home in Carolina as elsewhere in the Republic. No Chinese wall can now be tolerated. The South must be opened to the light of law and liberty, and this

session of Congress is relied upon to accomplish this
important work." Frederick Douglass[2]

"Out of the 'irrepressible conflict' between freedom
and slavery has grown one of far graver portent to
the nation and the world. Ignorance, poverty, inher-
ited barbarism, in that transition period took up the
conflict for equality of right and parity of authority,
against intelligence, wealth, experience, and the bitter
prejudice which centuries had engendered between
subject-black and dominant-white." [3]

The "eeevul Yankee carpetbaggers" narrative about Reconstruction
was pushed by the 20th-century Planter elite, and accepted with a
shrug by indifferent mainstream historians. They knew better; they
just didn't care enough to fight the obsessed neo-Confederates who
made it their life mission to peddle the "carpetbaggers" version of
Reconstruction. This farce became known as the Lost Cause Myth,
in which the noble, doomed Confederacy, which was only fighting

---

2. "Reconstruction" by Frederick Douglass in The Atlantic, December 1866.

3. *An Appeal to Caesar* by Albion Winegar Tourgée. (New York:
   Fords, Howard, and Hulbert, 1884).

for "states's rights," (not slavery, oh no, Heaven forfend!) was over-whelmed by hordes of Northern mercenaries.

This nonsense reached such a shameful peak that even Henry Wirz, the only Confederate officer to be hanged for war crimes—Hen-ry Wirz, commander of the Confederate death camp at Anderson-ville—had an obelisk erected in his honor, if that word can be used here, by the Daughters of the Confederacy.

American leaders knew the Lost Cause storyline was disingenuous, and that the Confederacy had been, as Grant said, "the worst cause for which men ever fought." But they allowed the myth to spread because, to simplify a bit, they were more interested in coopting the white south for contemporary political gain than in trying to help Black people trapped in the south.

This shows you how powerful the trope of picking your starting date can be. Andrews could not possibly defend the premise that 19th-century Americans were not obsessed with race. So she just skips that century—the one in which Reconstruction actually took place—and jumps to the present, contrasted with a nicely fuzzy, dim golden age, "...most of the 20th century..." For most of her intend-ed audience, this translates into "before all this woke stuff started." Which is a mischaracterization even of the 20th century, let alone the 19th, which was obsessed with race with a frankness and to a degree unimaginable even to the most woke 21st-century writers.

The Lost Cause Myth was itself a "reinterpretation of history," a very violent one. In 1865, when the war ended, nobody, north or south, tried to deny that the Confederacy had fought to keep "African slavery." That was obvious. Most Confederate survivors didn't even bother to deny it.

But these sore losers, who were unfortunately allowed to reoccupy their estates and resume their vile pamphleteering tradition, immediately began "reinterpreting" the war, arguing that they hadn't fought for slavery and that they hadn't really lost at all.

The reality of their defeat and shame was too much, and though the better of the ex-Confederates (like Longstreet) did their best to accept reality[4], and some took refuge in opium or whiskey, many others—the very worst of the lot—started rewriting the history of the war and its aftermath, hoping to win the narrative of the war even after losing the war itself.

Jubal Early, a lesser Confederate general, took the lead in promulgating the Lost Cause line. In essence, this posture was what Freud would call "Kettle Logic": "We didn't really lose and besides the only reason we lost is that you cheated."

Early wrote the first and biggest Lost Cause book, *Memoir of the Last Year of the War for Independence*. Early's polemic was published

---

4. "Men can't all think alike, and the trouble with the Southern people always has been that they won't tolerate any difference of opinion. If God Almighty had intended all men to think just alike, He might just as well have made but one man....My opinion is that the only true solution for Southern troubles is for the people to accept cordially and in good faith all the results of the war, including the reconstruction measures, the acts of Congress, negro suffrage, etc., and live up to them like men. If they would do this, and encourage Northern immigration, and treat all men fairly, whites and blacks, the troubles would soon be over, and in less than five years, the South would be in the enjoyment of greater prosperity than ever." Interview with correspondent from the Indianapolis Journal, September 24, 1874.

a year after the war ended, in Toronto, which was a very pro-Confederate city. Early was one of the great liars in history, unhinged and ridiculous. He spent the last year of the war losing a series of battles in the Shenandoah Valley and then tried to win those battles in post-war prose. Early's work resembles more than anything the pitiful self-justifications of the U.S. commander in Vietnam William Westmoreland. Westmoreleand, who was born in Spartanburg (!), South Carolina, spent his last years proving in self-justifying prose that he did not really lose to the Vietnamese Communists because they didn't play fair. It must be a Southern tradition or something.

Early took over a Confederate force that had been victorious in many battles to keep the Shenandoah. He was decisively beaten in a series of quick cavalry battles by a new Union commander, Phil Sheridan. Even Lee praised Sheridan's military ability when talking to Grant after Appomattox, and Lee made his feelings about Early quite clear by dismissing him from the Shenandoah command in the last months of the war.

The Union general J. W. Keifer, who'd played a big role in Early's humiliation in Virginia, wrote that it was Early's complete failure as a commander that made him such a vindictive, tireless pusher of the Lost Cause Myth:

"[Early's] misfortunes in the [Shenandoah] Valley, doubtless, had much to do with his continued implacable hatred to the Union."[5]

Early and his followers were the most ideologically driven revisionists in American history. I mean, it's not even close. You don't need to tell me about how annoying academic leftists can be. I have BEEN THERE with those defrocked Congregationalist clerics, believe me, and I have the scars to prove it.

But nobody could ever say that those buzzing little gnats were one-tenth as wrong, or as dishonest, or as influential as the swine who used the Lost Cause myth to reimpose serfdom on generations of African Americans.

More than 300,000 loyal American soldiers died horribly in the Civil War, with hundreds of thousands more maimed in ways that 19th-century medicine couldn't even grasp, let alone heal. What Jubal Early and his followers did was ensure that all that suffering would be for nothing, and the Old South would be born again after the war, in a more sly, cowardly form, to impose its obsession with race again, without the name of slavery but with all its cruelty and terror, so that the struggle had to be resumed again, and the Lost Cause Myth pulled down, slowly and against constant resistance.

---

5. *Slavery and Four Years of War, Vol. 1-2:* A Political History of Slavery in the United States Together With a Narrative of the Campaigns and Battles of the Civil War In Which the Author Took *Part: 1861-1865* by Joseph Warren Keifer. (first published 1900). June 2008, Kessinger Publishing.

The rest of Andrews's miserable article consists of an attempt to debunk W. E. B. Dubois's book *Black Reconstruction*, a takedown so inept that Andrews cites Dubois's assertion "I am going to tell this story as though Negroes were ordinary human beings" as if it somehow condemned him.

In fact, Dubois's take was exactly right, obviously right. Reconstruction was a struggle between those who wanted African-Americans in the south to be treated as if they were ordinary human beings, and those who insisted that they were less than human. How anyone, even this Helen Andrews, can fail to see that it was the Lost Cause advocates who "reinterpreted" the story to make it "...about race..." is beyond me.

Her attack on Reconstruction is so lacking in context that it makes me veer back toward the "ignorant" pole of the "Is she ignorant or just pretending to be?" question which is the most interesting one about her polemic. For example, she zooms in on a rather odd issue: how much the Mississippi legislature spent in the years when it took the radical, leftist step of allowing black people to serve in it. Here's her blockbuster exposé:

> "Perhaps the figures do not prove theft but they certainly suggest it. Between 1868 and 1872, the South Carolina legislature appropriated $200,000 for furniture; when auditors examined the State House in 1877, only $17,715 worth of furniture (in original prices) was found; in 1890, the whole House chamber was refurbished for $3,061. Expenditure on champagne and whiskey for the Columbia State House was $125,000 in a single year, equivalent to about $1.5 million today. Other states, such as Louisiana, saw

tenfold increases in their budgets relative to prewar
averages. Du Bois suggests this money might have
been 'spent carefully and honestly upon legitimate
and necessary matters of restoration and government.'
No one at the time was so naïve."

Well, that sure breaks the story wide open. What?! Rookie legislators from a deeply impoverished, oppressed minority engaging in
graft? Say it ain't so! Ah, just wait until Andrews hears about Tam-
many Hall. Or, God forbid, starts thinking a little harder and wonders
about how all-white legislatures in the antebellum south allocated
funds.

People like Andrews always get indignant when legislators who
came from poverty (or in this case, outright slavery) graft a little champagne for themselves "at the taxpayers' expense." Do they ever wonder
how those all-white, all-plantation-owner southern legislators of the
antebellum south bought their champagne? With the whip, Madam,
with the whip!

To Ivy League conservatives like Andrews, there's nothing wrong
with whipping your slaves into producing enough surplus value to
buy casks of Parisian champagne. That's not "corruption" because it
wasn't the government, it was private enterprise. Which BTW was
exactly what the Confederates said.

Bizarre, isn't it? This Yalie is actually arguing in 2021 that Dubois's
1935 book was wrong in tallying up the SC legislature's expenses. One
would almost suspect that she hasn't done any research of her own
and is simply looking for paragraphs of Dubois's text, a century late,
for cheap contemporary polemical purposes.

It's none of my business, really, but when I read the bio's of people
like Andrews and see their triple-dyed sanctity on parade, I get a little

shiver. I mean, at least the performative leftists are mostly atheists, which I guess allows them to strut a spiritual pride without fearing damnation, which seems very sinful to me. Andrews actually claims to be a believer, yet she lies like this about real, massive, protracted human suffering—and all to score a quick point with people who wouldn't know W. E. B. from Blanche Dubois.

Don't they teach you about Hell in Religious Studies at Yale?

Chapter Seventeen

Speaking of Monuments That Need Sledgehammers...

*T*he Lost Cause was an early-twentieth-century phenomenon and, *though some of the statues put up have been torn down, there's one in a little place called Andersonville that still endures. Andersonville is the best proof that there are no ghosts, because if anyplace deserved to be haunted, Andersonville would be more crowded than Mumbai.*

The next time you run into a Confederate apologist, just say one word: Andersonville. That should settle the argument. It won't, of course—but it should. Because there was no excuse for the abomination the Confederacy built in that little southwest Georgia town.

Most Civil War fans will know, vaguely, that there was a Confederate POW camp near Andersonville, and that it was an unpleasant place, and that many prisoners died. The truth is much, much worse. Andersonville was a death camp. No one will admit that, but if the stats came from a POW camp in any other country, we'd have no hesitation calling it what it was. This is part of a pattern we see over and over: "McClellan was not a traitor of course"; "Of course Andersonville was not meant to be a death camp"; "Of course John Pope's officers did not scheme to cause his defeat." These are all products of a childish refusal to accuse people we think of as fellow Americans of being what they were: murderous traitors.

Like a lot of death camps, Andersonville was a vindictive product of defeat. The prison camp there started up in February 1864, after almost three years of war—and about six months after it was clear, to everyone who counted in the South, that they were going to lose.

That made a huge difference in the way they treated their prisoners. This often happens; once an army faces final defeat, the urge to take revenge-in-advance on captured enemies always grows more powerful.

To explain the murderous hate that created Andersonville Prison, I'll have to zoom through the arc of the war, so apologies to those who'll find this old news.

For the first two years of the war, the Confederate armies won most of their battles, especially in the Eastern theater. So the Southern elite was relatively humane with captured Union soldiers. Of course, there are lots of exceptions to that generalization, especially on the fringes of the war, in places like Arkansas, Texas, Missouri, and anyplace the war devolved into irregular skirmishing and raids. In such places, no-quarter battles were very common. A lot of wounded and captured Union men were shot or bayoneted when their positions were overrun.

But you can still say that in the first two years of the war, most Union POWs who lived through the moment of surrender (always a very dangerous one in war) had a good chance of getting shunted, however brutally, to a minimally livable POW camp where they would be fed, however badly, and sheltered, however crudely. Most of these camps were in Virginia (15 of them just around Richmond), with others in commandeered Federal installations, or any other available structures (including a racetrack). Islands were ideal prisons, so Confederate authorities set up one of their biggest POW camps on Belle Isle in Northern Virginia.

Conditions in these early camps were bad, even by mid-19th-century standards, for several reasons. First, the Confederacy was a classic example of the over-mobilized war state, where everything went to the front and the bosses didn't much care about their own civilians, never mind prisoners. Second, there was always a disparity in hate between North and South. Few Northerners truly hated the Confederates, and all too many were quite indulgent toward them. But many if not most white Southerners truly hated the Northerners.

Still, the camps where Union POWs were held, though chaotic and badly supplied, didn't show the pattern of intentional abuse that came later in places like Andersonville.

Some camps were halfway decent; some were miserable. Treatment varied with status, as it does in every war. Captured Union officers were treated better than enlisted men, and "Americans"—native speakers of English who looked Anglo—were treated better than "foreigners" i.e., immigrant soldiers. The Confederates hated those "foreign mercenaries," because most pre-war migrants came to the cities of the North or the cheap farmlands of the Northwest, rather than the slave-run agricultural states of the South, leading to intense xenophobia in the Confederate ranks.

But the truly vindictive, murderous policies toward Union POWs came after the Summer of 1863, about seven months before the Andersonville POW camp started up.

This was because at the beginning of July, 1863, the Union won two huge victories, capturing Vicksburg, key Mississippi outpost for the South, and smashing Lee's invasion force at Gettysburg. After the Union victories at Gettysburg and Vicksburg, the South had no real hope of winning the war. The Mississippi River was in complete Union control from Minnesota down to the Gulf of Mexico, and after two tries (Antietam and Gettysburg), it was clear that the South could not successfully invade the North. So the Confederacy was doomed, as most classically trained officers understood. And not all Confederate officers were courteous, chivalrous losers as we've been led to believe by generations of Lost Cause tragedians. Many of the officers who had fallen into the less prestigious work of guarding prisoners were incompetent and vengeful.

The most appalling example of this vile species is General John H. Winder, commander of prisoner of war camps in the South throughout the Civil War until his death of a heart attack in February 1865. Winder had been a life-long failure and an embittered one, somehow failing even to make money as a Planter using enslaved labor on his father's plantation, and known only for his nasty temper and general incompetence in the pre-war officer corps. He ran true to course as a Confederate official, running the Richmond police force openly as a career opportunity for the "plug-uglies" imported from Baltimore purely to terrorize the population, though there wree constant demands for an investigation for this so-called police force, Winder refused to pursue any probes into the charges. He seemed to enjoy inflicting pain on the helpless, whether these were the civilians of Richmond or Union POWs. This makes it all the odder that there is a tradition in American Civil War historiography represented by Ezra J. Warner claiming that Winder must have done the best he could. This is simply nonsense and it represents a very pernicious tendency in Civil War historiography: claiming that even when the CSA elite committed horrors, they must have done so because they were confused or misled rather than because they were just plain evil. Warner says, "Warner adopted every means at his command to ensure that the [Union] prisoners received the same rations as did Confederate soldiers in the field." This is the sort of pernicious drivel that rises from a sense that Americans are born without original sin. There is every evidence that Winder was a purely evil man for his entire life and even one of his would-be defenders, John McElory admits that he heard Winder say that he was "killing off more Yankees than twenty regiments in Lee's army" and that Winder, late in the war, ordered his troops at Andersonville to use grapeshot to massacre the prisoners if Union troops got close enough to liberate them.

To me, the contortions to which mainstream historians go to redeem these irredeemable characters, is rather puzzling and suggests a bias that probably vitiates a great deal of work done on the war in general.

Henry Wirz was the crippled, broken wretch who ended up in command of the Anderson POW camp. He was hanged, deservedly so, but he was hardly worth hanging, a man of no importance. Winder, alas, was not hangable because he'd died before the war ended and, no doubt, was enjoying the delights of Hell long before Appomatox.

We owe much of our knowledge of conditions at Andersonville to a lucky survivor, John L. Ransom, who was captured in Tennessee and moved to Andersonville. What sets Ransom apart from most of the wretched survivors of this hellish death camp is that he managed to keep a diary of his experiences there.

Ransom was unequivocal in his hatred for Wirz and insistence that Wirz never showed any compassion whatever for those in his charge. He described him as "Certainly the worst man I ever saw." You might think Ransom, a POW, is a bit prejudiced here, since he's describing the boss of his POW camp, but no—Wirz was a monster by any standards. And BTW, it's not actually true that prisoners hate their captors right across the board. They can't afford to do that. Prisoners always make very sharp distinctions among the people who control them—which guard is a sadist, who's a softie, which will listen to pleading, who can be bribed, etc.

And the prisoners' verdict on Wirz was unanimous. As Ransom said, "There are a thousand men in [Andersonville] who would willingly die if they could kill him first."

Wirz had a muddled history before the war, but then so did many of those in both armies. He was an immigrant, born in Switzerland. He moved to Turin, where he went to prison for debt (hardly a monstrous crime) and was released on condition he leave the country (Piedmont-Savoy, that is). His wife didn't care to go with him; they divorced. He sailed to the U.S., bounced around both North and South, remarried, and ended up in Louisiana. So he enlisted with the Confederacy.

Wirz was maimed by the time he took command of Andersonville. The bones in his right arm had been smashed by a bullet (it's not clear when or where this happened). After that, Confederate authorities sent him to Europe to plead their cause. But Wirz got nowhere. He was a convict, a lower-middle-class nobody, and by all accounts a rather unpleasant person. He sailed home to the Confederacy and got the Col. Klink job: supervisor of a POW camp.

"Camp Sumter" (Andersonville) was a product of the late war, when bitterness became the norm in what was left of the Confederacy.

Sherman hadn't begun his famous march yet—hadn't even taken Atlanta—but his slice through the Confederate heartland was like a slow disembowelment, teased out for added humiliation. Sherman openly gloated about the South's fatal over-mobilization: "The South is hollow, all hollow, inside!"

That revelation humiliated the Confederacy, making an already very brutal state even more disposed to cruelty. Andersonville was the product of this late-war attitude. It was designed to torture and kill the men thrown into it, a 16-acre pen, just a pen, with only a slow trickle of swampy water flowing through it.

You would almost have to search to find a place so barren in Southwest Georgia, the region where the camp was located. SW Georgia is a very rich country. You can hardly get away from clean, flowing water. Vegetables grow even when neglected. Timber for building was free for the taking. And all that was purposely withheld from the POWS at Andersonville. Prisoners could have been allowed out in small groups chosen for their reliability. These small groups could have brought back enough sweet potatoes to keep everyone alive and relatively healthy, but that is not what the camp was for; it was meant to inflict suffering and slow death and it performed that function very well.

Wirz took to his job as camp commander with enthusiasm. John Ransom, already among the doomed inside the pen at Andersonville, describes his first sight of Wirz:

> "Captain Wirtz [sic] [came] inside today, and looked us over. Is not a very prepossessing looking chap. is about 35 or 40 years old, rather tall, and a little stoop shouldered; skin has a pale, white-livered look, with thin lips. Has a sneering sort of cast of countenance.

Makes a fellow feel as if he would like to go up and boot him."

There's a clear trend in the way prisoners were treated at Andersonville. From the start, they were penned up to die—but as the Confederacy weakened, the brutality increased. Forrest's attack on Fort Pillow, and the massacre of black soldiers after surrender, made a big impact even inside the pen at Andersonville, as Ransom reports:

"Fort Pillow prisoners tell some hard stories against the Confederacy at the treatment they received after their capture. They came here nearly starved to death, and a good many were wounded after their surrender."

Once thrown into the pen at Andersonville, prisoners lay on the open ground, which crawled with maggots to the point that the earth seemed to be moving. There were no latrines; the smell soon became so bad that new arrivals would faint when shoved inside.

Ransom describes the scene:

"Insects of all descriptions making their appearance, such as lizards, a worm four or five inches long, flea, maggots &c....New prisoners are made sick the first few hours of their arrival by the stench which pervades the prison. Old prisoners do not mind it so much, having become used to it. No visitors come near us anymore. Everybody sick, almost, with scurvy—an awful disease."

Scurvy was known to be easily prevented by adding fresh vegetables to prisoners' diet. Sweet potatoes were rotting in the fields of Georgia. But prisoners at Andersonville got one meal a day—grain-based gruel. The rains came, then the heat, but no shelter was available:

> "Laying on the ground so much has made sores on nearly everyone here, and in many cases gangrene sets in, and they are very bad off. Have many sores on my body...Today I saw a man with a bullet hole in his head, over an inch deep, and you could look down in it, and see maggots squirming around at the bottom. Such things are terrible, but...common."

Soon, the men's teeth get sore, then spongy, then begin falling out. Ransom:

> "My mouth getting sore from scurvy, and teeth loos e....Scurvy and dropsy are taking hold of the men."

Dropsy made their legs swell up even as their bodies wasted away from starvation. Ransom calls scurvy,

> "...about the most loathsome disease, and when dropsy takes hold with the scurvy, it is terrible. I have both diseases...my legs are swollen, but the cords are not contracted much, and I can still walk very well."

Ransom survived the camp, while almost all his friends died—but he was not so lucky with his legs. The "cords" in his right leg did contract, to the point that he came home with a very bad limp. Still, he was lucky.

Of the 45,000 men imprisoned in Andersonville, 13,000 died. That's an astronomical death rate, and despite what the neo-Confederate shills say, there was nothing even approaching it at any Northern POW camp. The death rate for Confederate POWs in the North was 12 percent, vs. 28 at Andersonville.

Andersonville was a piece of cleared land inside a huge forested area. Prison staff made excellent use of this free timber—but not for the benefit of their prisoners. All improvements were designed to make the camp escape-proof rather than liveable. Slave labor was brought in to dig a deep ditch around the pen, to make sure no tunnels could be dug. Artillery was emplaced, supplied with grape and cannister to mow down any attempt at mass escape. There was plenty of energy for these projects. But none to keep the skeletons inside alive.

Imagine this situation with American POWs held by any other power. They're stuck in an open pen for years, with no fresh food, no latrines, no fresh water, no shelter. That's a death camp. We'd call it that if the soldiers penning up our prisoners were German, or Japanese, or Russian, or Chinese. We're just too chicken to say it about English-speaking Confederates.

Wirz's superior actually admitted that killing off Union soldiers was the whole point of the camp. According to John McElroy, a survivor of Andersonville, General John Winder, a nasty specimen even by the standards of the Confederate officer corps, boasted that he was "killing off more Yankees than 20 regiments of Lee's army" by starving prisoners to death at Andersonville.

McElroy discovered a memo from Winder ordering the massacre of all prisoners if Union troops came within seven miles of the camp:

> *Headquarters Military Prison, Andersonville, Ga., July 27, 1864, Order Ko. 13.* The officers on duty and in charge of the Battery of Florida Artillery at the time will, upon receiving notice that the enemy has approached within seven miles of this post, open upon the Stockade with grapeshot, without reference to the situation beyond these lines of defense. *John H. Winder, Brigadier General Commanding.*

Winder escaped the noose by dying of a heart attack while sitting down to a hearty meal in 1865. Wirz, the fall guy on the spot—and totally deserving of a noose as well—became the only Confederate officer to hang for war crimes. At his trial, locals testified that there were always plenty of sweet potatoes and other fresh vegetables available, all through the months when POWs were being starved to death.

But Wirz still became a hero for Lost Cause fans at the beginning of the 20th century. So much so that the United Daughters of the Confederacy actually put up a monument to Wirz in Andersonville. Believe it or not, the big controversy was whether Andersonville should have the honor of celebrating this man, or whether the monument should be placed in Richmond, VA. They fought over it like they were trying to win the contract to host an Olympic Games.

Andersonville won out, and the obelisk went up. There are inscriptions on each of its four sides. They're worth quoting, to get the full insane bitterness of the Lost Cause sensibility:

North Side: When time shall have softened passion and prejudice, when reason shall have stripped the mask from misrepresentations, then justice, holding evenly her scales, will require much of past censures and praise to change places.—Jefferson Davis, Dec. 1888

South Side: Discharging his duty with such humanity as the harsh circumstances of the times, and the policy of the foe permitted Capt. Wirz became at last the victim of a misdirected popular clamor. He was arrested in the time of peace, while under the protection of parole, tried by a military commission of a service to which he did not belong, and condemned to ignominious death on charges of excessive cruelty to Federal prisoners. He indignantly spurned a pardon proffered on condition that he would incriminate President Davis and thus exonerate himself from charges of which both were innocent.

East Side: In memory of Captain Henry Wirz, C .S.A. born Zurich, Switzerland, 1822, sentenced to death and executed at Washington D.C. November 10, 1865. To rescue his name from the stigma attached to it by embittered prejudice this shaft is erected by the Georgia division, United Daughters of the Confederacy.

West Side: It is hard on our men held in southern
prisons not to exchange them, but it is humanity to
those left in the ranks to fight our battles. At this particular time to release all rebel prisoners would insure
Sherman's defeat and would compromise our safety
here.—Ulysses S. Grant, Aug. 18, 1864.

And in case you think nobody now would take this sort of craziness
seriously, here's a site where a neo-Confederate actually picks up on
the "West Side" inscription to blame none other than U. S. Grant for
intentionally starving 13,000 American POWs to death:

"Much of the responsibility for what happened at
Andersonville really fell on the shoulders of the leaders of the Union war effort, President Abraham Lincoln and General Ulysses S. Grant."

In August 2025, the monument to Wirz still stands in Andersonville. So if anybody wants to take a field trip to Andersonville, bring
your sledgehammer and we'll all pile into a few buses and turn that
obelisk into Confederate gravel.

Original Titles and Publication Details

C hapter 1. "Monitor and Merrimack, My Ironclad Gods." eX iledonline.com, April 23, 2011.

Chapter 2. "The Confederates Who Should Have Been Hanged." Pando.com, April 10, 2015.

Chapter 3. "The Diary of Adam Gurowski." Pando.com, May 22, 2013.

Chapter 4. "Yes, They Were Traitors: John Logan's Great Conspiracy." RWN Newsletter #121, January 9, 2022.

Chapter 5. "McClellan: 'Top of His Class at the Point." RWN Newsletter #130, November 19, 2022.

Chapter 6. "Poe-Land"—Poe at West Point." RWN Newsletter #115, June 18, 2018.

Chapter 7. "A Traitor Writes Home: McLellan's Harrison's Landing Letter." RWN Newsletter #131, December 2, 2022.

Chapter 8. "Worst Battle Speech Ever." RWN Newsletter #126, May 31, 2022.

Chapter 9. "Arm the Slaves? The Confederacy Would Rather Die." RWN Newsletter #128, August 5, 2022.

Chapter 10. "Turchin's Trial: The Shift to Hard War," RWN Newsletter #122, February 2, 2022.

Chapter 11. "Why Sherman Was Right to Burn Atlanta." NSFW-CORP, November 20, 2014.

Chapter 12. "Keep the Home Fires Burning: Civil War Arson, Spinal Tuberculosis, and the Confederate Army of Manhattan." RWN Newsletter #135, May 12, 2023.

Chapter 13. "Twain at War." RWN Newsletter #127, June 22, 2022.

Chapter 14. "Cherow's Big Pithed Plinth." RWN Newsletter #101, August 27, 2020.

Chapter 15. "Women's Civil War Journals." RWN Newsletter #141, December 23, 2024.

Chapter 16. "That Weird AmCom Article on Reconstruction." RWN Newsletter #120, December 16, 2021.

Chapter 17. "Speaking of Monuments That Need Sledgehammers." RWN Newsletter #55, August 31, 2017.

Acknowledgements

A s always, there are too many people to thank. First of all, the subscribers to Radio War Nerd who called repeatedly for a Civil War series, overcoming my odd fear that a war so sacred in my childhood could never be properly conveyed to a younger audience. To Mark Ames, my cohost on Radio War Nerd, who has a perfect sense of how to organize a valid project and make it work. To Paul Carr, who published several of these pieces. To Gabriel Uriarte, who proofread the manuscript and offered perceptive suggestions. To Dan for the title idea. To Katherine, who forced me to act like a responsible author against my nature. And to George Thomas, a good man who lived his life in pain, stuck with the Union, and who was snubbed in life and death by his snotty Virginia relatives.

About the Author

John Dolan is an American writer and broadcaster. Since 2015, he and journalist Mark Ames have co-hosted Radio War Nerd, a podcast casting light on military matters.

In 2002, he began writing military analysis under the pseudonym Gary Brecher, 'The War Nerd'. At that time he and Ames were editors for *The eXile*, a scurrilous English-language newspaper based in Moscow. He has since written dozens of articles with the 'War Nerd' byline and three books: *War Nerd* (2008), *The War Nerd Iliad* (2017), and *Erdogan Pizza* (2023).

Apart from the War Nerd identity, he has also published several books and many book reviews, poems, essays and academic articles. He is currently working on a sequel to his memoir *Pleasant Hell* (2007).

Subscribe to Radio War Nerd at https://www.patreon.com/c/radiowarnerd/home

Visit his homepage https://www.johndolanwrites.com/

Contact John at radiowarnerd@gmail.com